Helion & Company Limited
Unit 8 Amherst Business Centre
Budbrooke Road
Warwick
CV34 5WE
England
Tel. 01926 499 619
Email: info@helion.co.uk
Website: www.helion.co.uk
Twitter: @helionbooks
Visit our blog http://blog.helion.co.uk/

Text © Sanjay Badri-Maharaj 2022
Photographs © unless stated otherwise, by Sanjay Badri-Maharaj
Colour profiles © Goran Sudar and Tom Cooper 2022
Maps © Goran Sudar and Tom Cooper 2022

Designed and typeset by Farr out Publications, Wokingham, Berkshire
Cover design by Paul Hewitt, Battlefield Design (www.battlefield-design.co.uk)
Cover photo: (Photo by Deb Rana)

Every reasonable effort has been made to trace copyright holders and to obtain their permission for the use of copyright material. The author and publisher apologise for any errors or omissions in this work, and would be grateful if notified of any corrections that should be incorporated in future reprints or editions of this book.

ISBN 978-1-915070-58-6

British Library Cataloguing-in-Publication Data
A catalogue record for this book is available from the British Library

All rights reserved. No part of this publication may be reproduced, stored in a retrieval system, or transmitted, in any form, or by any means, electronic, mechanical, photocopying, recording or otherwise, without the express written consent of Helion & Company Limited.

We always welcome receiving book proposals from prospective authors.

CONTENTS

Abbreviations		2
1	The Indian Air Force – Nine Decades of Service	2
2	The Doctrine of the Indian Air Force	16
3	The Current State of the Indian Air Force	20
4	Modernizing the IAF's Fleet – Inductions and Upgrades	31
5	The IAF's Ground-Based Air Defence System	37
6	The Indian Air Force and Nuclear Weapons	48
7	India's Military Space Efforts: The IAF's Final Frontier	55
8	Conclusion	59
Bibliography		60
Notes		66
About the Author		70

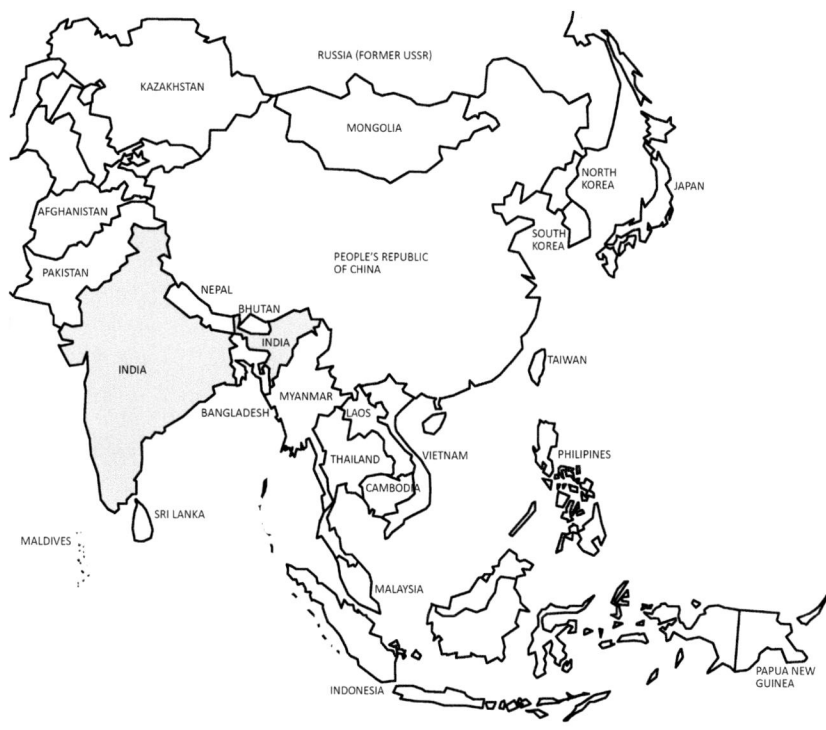

Note: In order to simplify the use of this book, all names, locations and geographic designations are as provided in *The Times World Atlas*, or other traditionally accepted major sources of reference, as of the time of described events.

ABBREVIATIONS

AAD	Advanced Air Defence	IAF	Indian Air Force
ADCC	Air Defence Control Centre	IACCCS	Integrated Air Command, Control and Communications System
ADGES	Air Defence Ground Environment System		
AEC	Atomic Energy Commission	INC	Indian National Congress
AEW	Airborne Early Warning	IRBM	Intermediate-Range Ballistic Missile
ALH	Advanced Light Helicopter	ICBM	Intercontinental Ballistic Missile
AMCA	Advanced Medium Combat Aircraft	LAC	Line of Actual Control
ARDE	Armament Research and Development Establishment	LCA	Light Combat Aircraft
		LCH	Light Combat Helicopter
ARM	Anti-Radiation Missile	LoC	Line of Control
BADZ	Base Air Defence Zone	MANPADS	Man Portable Air Defence System
BMD	Ballistic Missile Defence	MOP	Mobile Observation Post
BARC	Bhabha Atomic Research Centre	MRBM	Medium-Range Ballistic Missile
BEL	Bharat Electronics Limited	MRSAM	Medium-Range Surface to Air Missile
CBRN	Chemical, Biological, Radiological, Nuclear	QRSAM	Quick Reaction Surface to Air Missile
CEP	Circular Error Probable	SAM	Surface to Air Missile
CO	Commanding Officer	SLBM	Submarine-Launched Ballistic Missile
CRC	Control and Reporting Centre	SRBM	Short-Range Ballistic Missile
DRDO	Defence Research and Development Organisation	TEL	Transporter-Erector-Launcher
		UAV	Unmanned Aerial Vehicles
HAL	Hindustan Aeronautics Limited		

1
THE INDIAN AIR FORCE – NINE DECADES OF SERVICE

The Indian Air Force will celebrate its 90th anniversary on 8 October 2022. It has grown in capability by leaps and bounds in the past decades and while it is somewhat smaller than its peak strength in the late 1980s, it is a much more capable force with the capacity to project Indian power beyond the nation's borders. However, like the other armed services, it is in the throes of trying to match its modernisation efforts with resource constraints while simultaneously trying to achieve its doctrinal objectives.

The IAF has a proud history, dating back to its formation on 8 October 1932, with its first squadron being formed on 1 April 1933 with a small number of Westland Waptis. The Second World War showed a rapid growth in the force and its performance in that conflict provided invaluable assistance to the Allied war effort in South-East Asia and it was rewarded with the appellation 'Royal Indian Air Force'. Flying Hurricanes and Vengeances, the IAF/RIAF flew a large number of close support and reconnaissance sorties in support of the 14th Army, comprising troops from the Commonwealth, 70 percent of whom were from India.

The IAF, in the decades following independence grew at a steady pace, though its equipment choices and decisions in respect of force levels were always questionable. However, the IAF, while clearly in the shadow of the much larger Indian army, began to emerge as a force of its own, with a doctrine, a vision for where it wanted to be and a pathway to developing itself as an aerospace force, not just an

The aircraft that started it all – the Westland Wapti. (Author's photograph)

Hawker Tempest, the most powerful piston-engined fighter of the Indian Air Force of the late 1940s. (Author's photograph)

air force. However, its path is not and has not been a smooth one and nowhere is this more evident than its equipment mix, which varies considerably.

An Ad-Hoc Arsenal

The Indian Air Force started off operating British combat aircraft, with the Hawker Tempests and Supermarine Spitfires of the late 1940s being augmented with De Havilland Vampire jets. The venerable C-47 Dakota provided the main transport asset. The IAF moved slowly into the jet age as Vampires were augmented in the early 1950s with Dassault Ouragan fighters and later by Mystere IVA fighter-bombers. The Hawker Hunter entered IAF service in the late 1950s and would prove to be a mainstay of the IAF. Bomber assets were provided, initially, by a force of reassembled B-24 Liberators, and later, Canberras.

The IAF acquired the Folland Gnat in large numbers and as will be discussed below, sought to establish an indigenous aircraft industry, alongside licenced building of foreign designs. India produced a fighter-bomber in the form of the HAL Marut and a full range of trainers. However, the purchase and production of the MiG-21 saw the IAF adopt Soviet and Russian aircraft. This led to the acquisition of the MiG-23, MiG-27 and MiG-29 and the later purchase and manufacture of the Su-30MKI. Acquisitions of the Jaguar and Mirage 2000 rounded out India's ad-hoc inventory of combat aircraft, even as indigenisation floundered for want of investment. Yet, this very diversity and growth provide the foundation for the challenges faced by the IAF in the present. Maintenance, bloc obsolescence and the need to induct large numbers have had an impact on the IAF.

Efforts at Indigenisation

The Government of India sought to establish a domestic aviation industry. It began work on a basic piston-engine trainer to supplement and then supplant the Tiger Moths and Percival Prentice aircraft then in service. The result was the Hindustan HT-2 which served with distinction from 1953 until its retirement in 1990. Over 170 were built, with a dozen being used to form the Ghanaian Air Force in 1959.

By starting with a basic trainer, HAL (Hindustan Aeronautics Limited) had embarked upon its learning process in a sensible manner and intended to develop this core competency into an advanced trainer – the HT-11 – and an armed trainer – the HT-10 – which would have replaced the T-6 Harvard in the training roles. However, even at this early stage, short-sightedness combined with budgetary constraints conspired to stymie these plans. Neither aircraft progressed beyond the mock-up stage and a valuable learning process was ended prematurely.

Even at this early stage, HAL initiated some work in civil aircraft, with the HUL-26 Puspak trainer becoming a staple of Indian civil flying clubs following its first flight in 1958. An enlarged version, the HAOP-27 Krishak, formed the basis of army air observation flights until being replaced by Cheetah helicopters from the mid-1970s. The HAL HA-31 Basant crop-spraying aircraft had a limited production run (31 aircraft) but proved successful in service. For transport duties, the HS.748 Avro entered production in

One of 29 de Havilland Vampire NF. Mk 54 two-seat night fighters acquired by the IAF. India also acquired over 200 Vampire FB.Mk 52 single-seat fighter-bombers. (Author's photograph)

Starting in 1954, Indian Air Force acquired 104 Dassault Ouragan fighter-bombers from France. (Author's photograph)

Starting in 1957, India followed up with an order for 110 Dassault Mystere IVAs from France. (Author's photograph)

the 1960s and became progressively 'Indianised' with the survivors soldiering on in IAF and BSF service.

To this point, HAL's work had been unpretentious but essential. Building the foundation for a viable industry necessitates starting from the simplest of aircraft. However, the needs of the IAF required HAL to branch into combat aircraft manufacture at an early stage.

The first jet combat aircraft to be manufactured in India was the De Havilland Vampire in its FB.52 and T.55 variants. Under a licence granted in 1950, which included the Goblin 2 turbojet, India was able to replace its piston-engine fighters with jet aircraft in a systematic and low-risk manner while simultaneously building its aviation industry.

The years 1956 to 1959 were critical ones for the Indian aviation industry. In 1959, HAL received permission to proceed with the development of a basic jet trainer to replace the Vampire T.55s and the T-6 Harvard. In one of HAL's nearly unqualified successes, the resultant aircraft – the HJT-16 Kiran – first flew in 1964 and in a modified version, it continues to this day and is the IAF's basic trainer.

Simultaneously, HAL had laid the foundations for fighter production with a licence agreement for the Folland Gnat being signed in 1956. Dr Kurt Tank was engaged to begin work on designing the HF-24 Marut.

The Gnat, despite its British origins, became an Indian fighter. At its peak, HAL could build four Gnats per month and this diminutive fighter transformed the IAF's combat arm completely. HAL also received a licence to produce the Bristol Orpheus engine.

The HAL Ajeet, while intended to improve on the Gnat's performance, was only marginally successful; by 1975, the desired performance could only be achieved with a more powerful engine and more advanced avionics. While four squadrons of Ajeets served between 1975 and 1991, the type never achieved its potential.

It is the story of India's short-sightedness in engine development which wrecked the promising HF-24 Marut. The HF-24 was designed around the Orpheus B.Or.12 engine rated at 6,810 lbf (30.29 kN) dry and 8,170 lbf (36.34 kN) with afterburning which was being developed for the proposed Gnat Mk.2 interceptor and a NATO light-weight strike fighter. Unfortunately, the British authorities cancelled their requirement of the type and India, being unwilling to provide the modest sum required to complete development, was stuck with the non-afterburning Orpheus B.OR.2 Mk.703 rated at 4,850 lbf (21.57 kN) which ended up being used on the Marut. An Indian effort to fit afterburners to this engine resulted in between an

Hawker Hunter F.Mk 56 and sub-variants saw an extremely long service life with the IAF, and participation in multiple wars with Pakistan. (Author's photograph)

The venerable Douglas C-47 Dakota serves to this day with the BSF. (Author's photograph)

18 percent and 27 percent increase in thrust but the loss of the test aircraft, with Group Captain Suranjan Das, in 1970 ended this effort.

An attempt to re-power the Marut using Brander E-300 turbojets being developed for the Egyptian Helwan HA-300 fighter (an aircraft which in many ways was a supersonic Gnat-type light interceptor), each rated at 6,275 lbf (32.4 kN) dry and 10,582 lbf (47.2 kN) with afterburning, was potentially promising. However, form drag was considerable and while testing was satisfactory, the 1967 Six Day War ended this avenue of development.

It should be noted that while underpowered, the Marut was an excellent weapons platform and though somewhat short on range, its performance characteristics – even with the Mk.703 – were not dissimilar to contemporary types like the French Dassault Etendard IVM (which served until 1987) or even the Dassault Super Mystere B.2 (which continued in service until 1996 in Honduras). In contrast, the last Maruts left squadron service in 1985.

Despite some half-hearted efforts to find a suitable engine for the Marut, the IAF was never entirely supportive of the project. An attempt to integrate Adour turbofans (used in the Jaguars and Hawks) was confounded by an IAF demand that the thrust of the Adour be increased by 20 percent.

This decidedly unhelpful attitude was caused, at least in part, because the IAF's immediate requirements were being catered for by a substantial infusion of Soviet aircraft – the Su-7 for tactical strike and the MiG-21FL/M and MF variants. A very realistic and cost-effective proposal to create a strike-fighter based around the Marut airframe and the R-25 engine (the HF-25) received no sanction and while efforts to procure RB.199 turbofans were seriously considered for a Marut Mk.3 – the HF-73 – the project failed to materialise.

Ferdinand Brander, designer of the E-300, was far more blunt and firmly believed that the failure of India to develop an E-300 powered Marut was down to Soviet pressure and the desire of the USSR to sell MiG-21s and the licence to manufacture them in India.

As for HAL, its design expertise atrophied and initiative was discouraged. Its HPT-32 Deepak trainer was, until recently, its last success and even then HAL's upgrade of the type into the HTT-34 received no encouragement. Its efforts to replace the type with the HTT-35 – seen in mock-up form at Avia India '93 – also met with

Acquired when the USA had a crucial interest in helping India in its struggle against the People's Republic of China, the IAF's transport component was re-equipped with US-made Fairchild C-119 Packet transports. While piston-engined, they received Orpheus Jetpacks to bolster their 'hot and high' performance. (Author's photograph)

India's first indigenous military aircraft – the HT-2. (Author's photograph)

HAL HF-24 Marut. (Author's photograph)

no support. It must have been particularly galling for HAL to then see the IAF go in for the purchase of 75 PC-7 Mk.2 trainers which were very similar to their proposed HTT-35.

The Era of Licence Manufacture – A HAL-Assembled Air Force

HAL has invested enormous efforts into phased indigenisation of licence-manufactured products. This has come at considerable cost, but it has meant that licence-production in India is not mere assembly but involves a progressive increase in indigenous content. HAL has produced hundreds of combat aircraft and helicopters under licence and has achieved a high degree of indigenisation in these projects

HAL has also consistently failed to achieve indigenisation targets set for the Su-30MKI programme with raw material production of the aircraft beginning later than expected; even now the import content by value for the type is at 40 percent. Indigenisation levels have been at best modest for HAL's licence-manufactured products with even long-produced items such as the BAE Hawk and the Dornier Do-228 having a disproportionate import content by value – 60 percent in the case of the Do-228 and 58 percent in the case of the Hawk. Whether this was due to poor contract negotiation or is a failing of HAL is debatable, but it is undeniable that HAL's licence-produced aircraft have a relatively high import content by value which in part contributes to their high cost compared to direct imports.

HAL produced, and produces, MiG-21, MiG-27 and Su-30MKI aircraft under licence from the USSR/Russia, Jaguars and Hawks from the UK, Dornier Do-228 from Germany and Chetak and Cheetah helicopters from France. The degree of indigenisation achieved is broken down into indigenisation by content and indigenisation by value. By these measures, in the past, India achieved 90 percent indigenisation by content of the Chetak (72 percent of its engine), 88 percent of the Jaguar (84 percent of its engine) and over 96 percent of the MiG-21bis engine, again by content.

In more recent times, HAL has achieved a 75 percent indigenisation by content of the Su-30MKI (60 percent by value), 72 percent indigenisation by content of the BAE Hawk (42 percent

Sukhoi Su-7BMK was the last jet imported in large numbers from abroad without licensed production in India. (Author's photograph)

Type 77 or MiG-21FL – one of the first combat aircraft manufactured under license in India. (Author's photograph)

Polish-made TS.11 Iskra served as a trainer for decades, and then as an anti-UAV platform during the 2002–03 confrontation with Pakistan. (Author's photograph)

by value) and 73 percent by content of the Do-228 (40 percent by value). Contractual decisions as to indigenisation levels by content and value are decided by the priority to be given to the project and the timelines involved.

The IAF, therefore, built its strength on the basis of an extensive licence-production programme by HAL. This has, as noted, involved large-scale indigenisation efforts but not of Indian designs. The IAF was therefore able to build up a substantial squadron strength but the roots for later trouble were laid.

Growth and Baptism by Fire

The IAF has been involved in three major international conflicts between India and Pakistan – the wars of 1948, 1965 and 1971. It was also involved, albeit not in a combat role, in the 1962 war between India and China. Each of these wars, and the IAF operations therein, warrants its own study and a detailed analysis of them will not be undertaken in this book. Rather, a brief synopsis will be given, as this book deals with the IAF as it is today, which owes much to the IAF of the 1980s and 1990s rather than its earlier configurations.

The partition of India in 1947 saw the force divided between India and Pakistan and shortly thereafter, by 1948, the countries and their respective air forces found themselves at war over the then Princely State of Kashmir. This war saw air power being used for close air support as well as for transport of troops. Both air forces, small at the time, displayed initiative and determination but were supporting players to their respective ground forces.

The RIAF, flying eight squadrons of Tempests and one of DC-3s, augmented by Spitfires and Harvards, operating with armament from training squadrons, was able to provide effective reconnaissance and close air support to the Indian army. Their effort included the expenditure of 244 x 1,000lb bombs, 1,100 x 500lb bombs, 5334 3-inch rocket projectiles and 458,319 rounds of 20mm cannon ammunition. The Dakotas of No. 12 squadron delivered 3.5 million pounds of supplies, moved 4,000 troops and evacuated 10,000 refugees and 1,000 casualties in a four-month period between January and April 1948. As makeshift bombers, the Dakotas delivered some 40,000 pounds of bombs.[1]

The decades of the 1950s saw India grappling with innumerable challenges – social, economic and political – and while the IAF expanded and acquired new equipment, it suffered from an uneven expansion programme with aircraft in bizarre situations, such as where the IAF flew modern jet aircraft alongside piston engine fighters. The IAF was not immune from the reduction in defence expenditure and its logistics and ability to support its fleet was severely compromised for much of the 1950s and 1960s. The acquisition of new aircraft was not matched by an equivalent improvement in infrastructure or ordnance and the concept of operations for the force was not fully matured when the IAF found itself facing another challenge, this time on the China border, for which neither it nor the army was prepared.

The Sino-India war of 1962 was India's only post-Independence military defeat and it saw an extremely poorly equipped Indian Army, badly deployed and ill-supported, thrown into action against a well-motivated and innovative Chinese military. Neither side used air power in an offensive mode, though the IAF was prepared to support the Indian Army and could have, if allowed, turned the tide of the conflict. However, the political decision not to use air power was taken, partly under US pressure, and the IAF was left to play a supporting role to the Indian Army, keeping troops resupplied and ferrying men and equipment as the disaster unfolded.

In this supporting capacity, the IAF's transports – Caribous, C-119 Packets, An-12s, Dakotas and even Il-14s – were all pressed into service and delivered thousands of tonnes of supplies, under arduous conditions and on semi-prepared landing fields. The IAF transport force also airlifted two AMX-13 tanks which were to prove pivotal in the defence of Chushul and its vital airfield. Moreover, the IAF's fledgling helicopter fleet – with Bell 47s and Mi-4 helicopters – performed well in nearly impossible conditions and proved invaluable for casualty evacuation and the support of remote units, but there were never enough machines to suffice.[2]

By the time the 1965 India-Pakistan war broke out, the IAF had some 26 combat squadrons, comprising of a handful of MiG-21s operating alongside a large force of Gnat, Hunter and Mystere fighters, vintage Ouragan and Vampire fighter-bombers, Canberra and B-24 Liberator bombers, supported by an eclectic mix of transports with An-12s operating alongside Otters, Caribous, Dakotas and Packets. The force operated a mix of armed Vampire and Harvard trainers while Bell 47 and Mi-4s, along with some Alouettes, made up the helicopter force. In addition, the first squadrons of SA-2 SAMs were in service, supplemented by Bofors L-60 AAA guns operated by the Indian Army. In combat, however, against the better-organised and better-equipped Pakistan Air Force and its 17 combat squadrons, the IAF had mixed success. The IAF flew some 4,000 sorties, losing 59 aircraft in combat, 35 of which were lost on the ground to Pakistani bombing raids and 12 of them being lost in the East. Pakistan's losses were lower, but were anywhere between 25 and 43 aircraft lost in combat over 2,500 sorties.[3]

The IAF emerged chastened from this experience and six years later, when the two countries went to war in 1971, the IAF was able to achieve a decisive victory. Now a 39-combat squadron force with a decided Soviet influence – MiG-21s and Su-7s supplementing Gnats, Hunters, Maruts and the surviving Mysteres – the IAF was able to establish both air superiority and provide close air support. The IAF also conducted a major airborne operation, with the 50 Indian Para Brigade airdropping a full battalion. In addition, its Mi-4 helicopters participated in a highly effective heli-bridge, moving troops across rivers. In all, the IAF flew some 11,549 sorties, losing 56 aircraft in combat, claiming 75 Pakistani aircraft lost either in combat or captured in the East when Pakistan's forces surrendered. The IAF thus played a major role in India's victory in 1971.[4]

The Strength of the Air Force – Peaking in the 1980s … Declining Thereafter

The decade starting in 1980 was to become the most influential in the IAF's history. The growth, equipment mix and the eventual size of the force during that decade was to have an impact on the subsequent decades of the IAF, with the consequences of this period being felt to the present day as the IAF still grapples with equipment choices and a fleet size that was largely created in that decade. Maintaining this strength has proved a challenge.

The 1980s saw a major increase in the IAF's force levels with MiG-21s being inducted in ever-larger quantities. The IAF was able to replace most of its Folland Gnat squadrons with MiG-21s of various types and its HF-24 Marut squadrons were replaced with MiG-23BNs. However, the IAF's bomber force stagnated around the Canberra and while the Jaguar was inducted as a Deep Penetration Strike Aircraft (licence-produced by Hindustan Aeronautics Limited), the IAF never invested in a long-range bomber. The Su-7 was replaced by a combination of MiG-23BNs and MiG-27s with HAL undertaking licence production of the latter type.

Acquired to counter the General Dynamics F-16A/B Fighting Falcon of the Pakistan Air Force, MiG-23MFs served with Nos. 223 and 224 Squadrons, IAF. (Author's photograph)

The IAF fighter force suffered by comparison to the attention paid to the tactical and semi-strategic strike units. The MiG-21 received a minimal upgrade to its armament when R.550 Magic-1 and Magic-2 missiles were integrated to augment and later replace the ageing and decidedly unreliable K-13 (AA-2) Atoll missiles. In response to the Pakistani acquisition of F-16s, the IAF procured two squadrons of MiG-23MF fighters armed with R-23 and R-60 missiles, the latter type being retrofitted and integrated with the MiG-21 fleet. Even in the 1980s, the R-23 was at best of modest quality and was phased out rather early.

The biggest boost to the quality of the IAF's fighter force, although not in the quantities needed – an issue that bedevils the IAF to date – came in the form of the acquisition of two squadrons each of the Mirage 2000 and the MiG-29. The MiG-29 came with an armament that was completely optimised for interception, consisting solely of R-27 and R-60 missiles. The latter would be augmented and replaced by R-73s. The Mirage 2000s initially came without their Super 530D BVR missiles but these were later delivered, augmenting R.550 Magic-2 missiles and Belouga cluster munitions, conferring an excellent multirole capability.

The IAF entered the 1980s operating a SAM force of SA-2 missiles, which peaked at 21 squadrons. However, the increasing unreliability of the system and its ineffectiveness against low-altitude intruders meant that from 1981, the IAF began to acquire the Pechora SA-3 system from the USSR. These SAMs would supplement and later completely supplant the SA-2 in IAF service, with the last SA-2 systems being formally deactivated and phased out in 1991–92. The IAF would later acquire OSA-AKM (SA-8) systems to augment the Pechoras which eventually reached a strength of some 31 squadrons, providing India with its main SAM defence.

The Indian Air Force of 1988 was perhaps close to the peak of the service's combat squadron strength. Two squadrons of newly inducted Mirage 2000s – Nos. 1 and 7 – along with another two of MiG-29s – Nos. 28 and 47 – formed the cutting edge of the IAF's interception capability.[5] The Mirage 2000s, however, were still in the progress of receiving their complement of Super 530D missiles and Belouga submunition systems. The MiG-29s were initially armed with R-23 and R-60 air-to-air missiles but the R-23s were in the process of being replaced by vastly superior R-27 systems in both their SARH and IR guided variants.

Augmenting the MiG-29 and Mirage 2000 were three squadrons of Jaguar fighter-bombers – Nos. 5, 14 and 27 squadrons. The former Canberra squadrons – No. 6 and No. 16 – were earmarked for re-equipment and this was subsequently accomplished. The Jaguars were the most potent strike assets in the IAF at that time.[6]

Six squadrons flew the MiG-23. Four of these were tactical strike squadrons with MiG-23BNs and two more were interceptor squadrons with MiG-23MFs. The latter type, with R-23 and R-60 missiles, was inducted as an interim response to the Pakistan Air Force inducting F-16s and were deemed to be useful but modest assets. The MiG-23BN was an effective tactical strike platform, replacing the Su-7 in that role.[7]

The MiG-27, which was now being manufactured by Hindustan Aeronautics Limited alongside the Jaguar, had begun to equip No. 9 and No. 222 squadrons and would eventually equip no fewer than six Indian Air Force squadrons. Among the items of ordnance obtained for the MiG-27 fleet were Kh-25, Kh-25MP, Kh-29L and Kh-23 missiles.

The MiG-21, in its FL, M, MF and bis versions (production of the latter only ceasing at HAL in 1987) equipped no fewer than 19 squadrons – Nos. 3, 4, 8, 15, 17, 21, 23, 24, 26, 29, 40, 32, 37, 40, 45, 51, 52, 101 and 108. While functioning as multirole platforms with unguided rockets and bombs of up to 500kg – with the MF also capable of using the X-66 missile – the MiG-21 was primarily an interceptor and a mix of K-13, R-60 and R.550 air-to-air missiles were carried.

Alongside these supersonic combat assets were four squadrons of vintage platforms. A single squadron – No. 20 – continued to fly the Hawker Hunter (which also served as an advanced trainer) – while three more – Nos. 2, 22 and 28 – continued to fly the HAL Ajeet in the close support role. These aircraft were of limited capability but were all capable of carrying unguided rockets and bombs in the close support role. However, there was some work done in integrating the R-60 missile with the Hawker Hunter.[8]

Reconnaissance and electronic warfare assets rested on No. 35 Squadron with MiG-21s; Canberras with No. 106 Squadron using the Canberra PR.57; and No. 102 Squadron using the MiG-25R. The latter two types were to provide invaluable strategic reconnaissance information over the course of several decades.

The IAF's combat assets included some 38 SAM squadrons – 30 using a mix of SA-2 and SA-3 and eight using OSA-AKM (SA-8b) systems. As noted, the IAF SAM inventory eventually replaced all SA-2 squadrons, which peaked at a strength of 21 squadrons, with SA-3s. The latter type, equipping 31 squadrons, continues to form

One of eight MiG-25RB supersonic reconnaissance jets that were operated by No. 102 Squadron. (Author's photograph)

the backbone of the IAF SAM force to date though it is increasingly viewed as being obsolete, despite upgrades aimed at keeping them viable against present threats.[9]

Combat assets also included four helicopter units – one (125HU) with Mi-25 gunships and three of Chetaks, one of which was equipped with SS.11B1 anti-tank missiles. The Mi-25s and SS.11 armed Chetaks saw extensive service in Sri Lanka with the former even dropping 250kg bombs in addition to its more usual 57mm rockets and AT-2 'Swatter' anti-tank missiles (plus its 12.7mm rotary machine gun) while the Chetaks and their SS.11B1s were deployed for 'precision' strikes. Four additional units used Cheetahs for liaison and observation duties and for light transport. The Chetak and Cheetah, despite being poorly equipped for the reconnaissance role, have proved to be very reliable platforms and continue in service. The IAF and Army Aviation Corps armed several Chetaks and Cheetahs with 7.62mm MAG machine guns during Operation Pawan and viewed this particular modification as being very effective and surprisingly versatile for supporting ground troops as well as deterring attacks on troop transport helicopters.[10]

Other helicopter assets for the transport role included one unit with Mi-26 heavy-lift helicopters (126HU) and eight with a combination of Mi-8/-17 helicopters which could also be armed with 57mm rocket pods. The Mi-8/-17 units were to prove their worth during Operation Pawan and remained a useful asset for proximate interventions.[11] The Mi-8/-17 replaced the Mi-4 helicopter in IAF service and have proven to be the most effective medium-lift helicopters operated by the IAF since its inception.

In 1988, their potential was still being realised with more and more soldiers becoming used to heliborne operations and new tactics being devised to make effective use of these assets on the subcontinental battlefield. However, their numbers were as yet relatively modest and would require judicious use to achieve the greatest effect. This was not always easy as the IAF as yet lacked full night-operations capability for the type and this led to complications. It should also be noted that the IAF's helicopters played and continue to play a critical role in supporting the Indian Army on the Siachen Glacier where air supply is the only means of sustaining the Indian brigade located at extreme altitudes.

The IAF also maintained a large transport fleet, largely of tactical transports but including some significant strategic transport assets. Rather surprisingly, two squadrons of vintage platforms still soldiered on. One squadron still flew the DHC-3 Otter which used its STOL capabilities to great effect. Another squadron (No. 33) may have still operated a few DHC-4 Caribous which, though nominally withdrawn in 1986, might still have had a presence in IAF service.[12]

The HAL-built HS.748 filled a multiplicity of tasks with some sixty aircraft being in IAF service, and they also aided in the carriage of freight, the transport of personnel and liaison tasks. Though not designed as a military transport, the type performed good service for the IAF and remains in service to date, despite its evident shortcomings.[13]

From 1984, the Indian Air Force began receiving An-32 tactical transports and their induction into some six squadrons was able to completely transform the IAF's airlift capabilities with their enhanced payload and troop transport capabilities. They were used extensively to support Indian troops during Operation Pawan and the six squadrons of the type form the backbone of India's tactical transport fleet to this day. The An-32 was reportedly specially aimed at the Indian market as the An-26 was deemed by the IAF to have a wholly inadequate performance from 'hot and high' airfields.[14]

Strategic airlift in 1988 was provided by one squadron of An-12 (No. 25) and one of Il-76 (No. 44) heavy transports. Sufficient Il-76 aircraft would be obtained to allow the conversion of one flight of No. 25 squadron (the other operating An-32s) but in 1988, the An-12 and Il-76 reigned as the IAF's long-range transport assets and were to prove decisive during Operation Cactus.[15] The An-12s were adapted for use as makeshift bombers during the 1971 war and were reputed to be very effective in that role.

It should also be noted that many Indian combat types – MiG-21s, MiG-27s and Jaguars, its training aircraft (the HPT-32, HT-2 and HJT-16) as well the Chetak and Cheetah helicopters – alongside the HS-748 transport were being manufactured either under licence or, as indigenous designs by Hindustan Aeronautics Limited (HAL). While there were undoubtedly problems with HAL's production schedules and quality, its effort contributed enormously to the growth and modernisation of the IAF between the 1970s and 1980s.

Notable Operations

Operation Meghdoot, undertaken in 1984 in the Siachen Glacier – a region which is still held by Indian troops – required extensive support from the IAF. The IAF operated Air Observation Post squadrons – Air OP squadrons – which carried out observation tasks for Indian artillery. These units were equipped with a mix of Cheetah and Chetak helicopters – licence-built versions of the Alouette II and Alouette III respectively. The Cheetah, in particular, was to prove of immense value owing to its incredible ability to operate at high altitudes, albeit with very limited payloads. Initially, a force of six Cheetahs and two Mi-8 helicopters was sufficient for the period 1984–1985 but as the years and decades passed, it became clear that a greater number of helicopters of both light and medium types needed to be deployed in theatre and this expanded the IAF's involvement.

The IAF was heavily involved in supporting the Indian Army in Sri Lanka between 1987 and 1990. The Indian Peace Keeping Force (IPKF) was India's largest out-of-country military deployment and remains to this day a topic of some unease within India owing to the somewhat rushed and ill-conceived deployment of personnel as well as the unrealistic assumptions upon which its intervention was based.

The IPKF was the Indian military contingent, as part of a broader political objective, and was tasked with performing a major peacekeeping operation in Sri Lanka between 1987 and 1990 in pursuit of a political settlement between the Sri Lankan Government and Tamil political forces and insurgents. The IPKF was formed under the aegis of the 1987 Indo-Sri Lankan Accord that aimed to end the Sri Lankan Civil War between the Sri Lankan government and heavily armed Sri Lankan Tamil insurgents, most notably the Liberation Tigers of Tamil Eelam (LTTE).[16]

The tasks were assigned as per the terms of the Indo-Sri Lankan Accord, which was very much the brainchild of the then Indian Prime Minister Rajiv Gandhi. The impetus for the accord and the IPKF was the escalation of the ethnic conflict in Sri Lanka between the Sinhalese-dominated government and the Tamil minority, and a massive influx of Tamil refugees to India. Rajiv Gandhi took, in partnership with a much more reluctant Sri Lankan President J.R. Jayawardene, the step of pushing through the accord and establishing and deploying the IPKF.

The principal objectives of the IPKF was to disarm all Tamil insurgent groups, including the LTTE, and to establish a Tamil police force. The disarming of the insurgent groups was to be followed by the formation of an Interim Administrative Council as expeditiously as possible.

However, within a few months of their arrival, the IPKF, after some limited levels of disarmament by the insurgents, became engaged in heavy combat with the LTTE trying to enforce the terms of the accord. The LTTE tried to dominate the Interim Administrative Council, and refused to disarm. This latter aspect was a pre-condition to the establishment of peace and the enforcement of a political settlement on the island and threatened to wreck the tentative peace accord.[17]

At considerable cost, the IPKF established a level of military dominance in the affected regions, but it never succeeded in eliminating the LTTE. A change in government in India, pressure from Tamil political parties in India as well as an increasing feeling that the effort in Sri Lanka was not worth the cost led to a cooling of Indian enthusiasm for armed involvement in the conflict. This was further complicated by political machinations between a new Sri Lankan government and the LTTE.[18]

The Indian Air Force flew no fewer than 48,752 sorties without loss during this operation. While no combat aircraft were involved, the IAF deployed its Mi-25 attack helicopters and SS.11 armed Chetak helicopters against the LTTE and its transport helicopters and aircraft were employed in transporting and sustaining troops. Units deployed were as follows:[19]

No. 33 Squadron – Antonov An-32s
No 109 and No. 119 Helicopter Units – Mil Mi-8 helicopters
No. 125 HU – Mil Mi-24s
No. 664 AOP Squadron – Chetak and Cheetah

Operation Cactus – 1988

India's most dramatic intervention was in 1988 when it staged a daring operation in the Maldives. Operation Cactus was spearheaded by two Il-76 transports of No.44 Squadron IAF. In total, the two Il-76s carried 316 troops, one jeep-mounted 106mm M40A1 recoilless rifle and one communications jeep as well as additional equipment and first line ammunition for the troops.[20] Within 24 hours, India had deployed 1,650 personnel who, in addition to their normal combat package, also brought with them some four jeeps fitted with 106mm M40A1 recoilless rifles, seven 84mm Carl Gustav rocket launchers, seven medium machine guns, seven mortars – both 51mm and possibly 81mm, and, at least according to one account, 11 RPGs, which were not a standard weapon in the Indian military, and 16 other vehicles.[21]

Supporting this was an equally impressive air effort, given the circumstances and other commitments. Some four Il-76s were committed to the operation, joined later by a fifth. These were used in support of the 50th Para Brigade but also in supporting the four Mirage 2000H aircraft that were deployed to Trivandrum. These were augmented by some 25 of the 30 available An-32 transports, four Mi-8 helicopters and, rather surprisingly, at least two Canberra bombers which were deployed to Trivandrum along with the Mi-8s and Mirage 2000s.[22]

The IAF closed the decade of the 1980s with a large establishment and squadron strength. However, there were very substantial deficiencies that were to haunt the IAF. The MiG-21 and MiG-23 squadrons had an excessively high attrition rate and the IAF failed to procure an advanced jet trainer, using the manifestly unsuitable MiG-21FLs and Hawker Hunters in that role.

Furthermore, in its very expansion, the roots of its challenges were laid. The IAF was operating a bloated force of MiG-21s. These, even by the early 1980s, were increasingly dated combat aircraft that at some point would present the problem of bloc obsolescence. Upgrade plans for the MiG-21 were hampered by a lack of suitable weapon systems until the 1990s, and even then, only a small proportion of the vast fleet was to receive this attention. Similarly, the IAF's large SAM force was already showing signs of becoming obsolete by the end of the 1980s yet upgrades and overhauls allowed it to survive to the present day. The IAF was curiously reticent about embracing some newer technologies with AEW platforms, smart munitions and even BVR weapons being obtained at a modest pace. Even electronic countermeasures, while present, were not given the level of priority necessary for the IAF to face the coming decades without facing multiple challenges in respect of its capabilities.

The 1990s – Modernisation and Declining Strength

By 1999 the Indian Air Force was – then as now – the fourth largest in the world; however, it had begun to show the first stages of a numerical decline, though a qualitative improvement, in its combat squadrons. The transport, helicopter and training fleets remained largely unchanged, though older transports were phased out and the Mi-17 increasingly supplanted the Mi-8.

The force operated no fewer than sixteen squadrons of strike aircraft. These comprised of five Jaguar squadrons, eight MiG-27M squadrons and three MiG-23BN squadrons – numbering well over three hundred aircraft.[23]

The MiG-23BN squadrons were earmarked for the Offensive Air Support (OAS) alongside the MiG-27 squadrons. However, the Indian MiG-27Ms are fitted with a RSBN-6S navigation system which is associated with the nuclear strike role. Mikoyan has openly acknowledged that the MiG-27 is nuclear capable.[24] The Jaguars were also known to be capable of carrying nuclear weapons. The 'toss-up' technique used to deliver nuclear weapons was practiced by IAF aircraft during exercise 'Hammerblow' in 1988.[25]

It should be noted that even the vast fleet of Indian MiG-21 combat aircraft was multi-tasked, with several squadrons being deployed with unguided munitions for the strike role. These aircraft would have a very limited range and payload and as such cannot be seriously considered as candidates for anything more than a tactical OAS support role. The then very comprehensive upgrade programme for the MiG-21bis aircraft of the Indian Air Force which culminated in the MiG-21 Bison added little to their ability to deliver air-to-surface weapons. Lacking an effective all-weather nav/attack system, the MiG-21 fleet, particularly the MiG-21M fleet, was limited to using unguided 57mm rockets and 250kg bombs with 240mm S-24 rockets providing their heavier armament. The MiG-21MF fleet did have the ability to launch Kh-23 missiles but it is unclear if any of those missiles were obtained by the 1990s.

The 1996 purchase of Sukhoi Su-30MKI aircraft and the technology transfer agreement to enable their manufacture in India was highly significant. However, since the Su-30 was originally designed as a long-range interceptor, in 1999 it was unclear as to how the Indian Air Force intended to use them. The Su-30 can, however, carry a very heavy weapons load – including nuclear weapons – over a very long range.

Despite the large number of aircraft at its disposal, the IAF was peculiarly underequipped for modern strike operations. While the MiG-27s were purchased with ordnance that included Kh-25MP, Kh-25, Kh-23 and Kh-29L missiles, the number of these missiles procured was relatively small in number. The aircraft had no capability, at the time, to deliver laser-guided bombs and were largely deployed with unguided munitions such as 57mm, 68mm and 240mm unguided rockets and 250kg and 500kg unguided bombs.

Of interest is the fact that both the MiG-27s and MiG-23BN fleets operated much the same unguided ordnance which included BAP100 runway denial munitions. The MiG-23BN, despite being in service in substantial numbers, did not have any stand-off munitions capability and was employed solely with unguided munitions and no upgrades to either the MiG-23BN or MiG-27 fleets had been undertaken at the time with one sole exception – some of the aircraft were finally fitted with chaff/flare dispensers.

The Jaguar fleet comprised the core of India's deep penetration strike assets and with its DARIN Nav/attack system, the Jaguar excelled at delivering unguided munitions with great accuracy. Like the MiG-23BN fleet, and like most of the MiG-27 fleet, India's Jaguars lacked stand-off munitions and integration with laser-guided bombs although India had been moving towards integrating such weapons.

The induction of the Su-30MKI saw the IAF obtaining Kh-29TE, Kh-31 and Kh-59 missiles. However, the most potent strike asset for precision work in the IAF was the Mirage 2000. Already armed with Belouga submunitions, the Mirage 2000 fleet also easily accommodated Paveway modifications and designator pods for laser-guided bomb delivery and had demonstrated the same. The ATLIS designator pod and Matra 1,000kg laser-guided bombs delivered with the aircraft were deemed to be extremely effective but were also viewed as exceedingly expensive so were sparingly used.

The Sukhoi Su-30s were primarily used as long-range interceptors in the 1990s, capable of intercepting targets at ranges exceeding 120km. India's interceptors were then equipped with a mix of French and Russian air-to-air missiles. All aircraft were cleared to launch R-60 (AA-8) and Magic R.550 short-range missiles while the MiG-29 and Su-30 were cleared to launch R-73 (AA-11) and R-27 (AA-10) with a limited number of R-77 (AA-12) missiles being in service by the end of the 1990s. The Mirage 2000 used R.550 Magic-2 missiles as well as the Matra Super 530D. The MiG-21-FL/-M/-MF and bis used the R.550 and the R.60 as their primary armament while the single MiG-23MF squadron used R.60s and R-23 missiles.

Kargil 1999[26]

The end of the decade saw the IAF being committed to Operation Safed Sagar where it conducted operations in support of the Indian Army's efforts to dislodge Pakistani intruders from the Kargil sector in 1999. The IAF deployed some 150 combat aircraft in the theatre though not all, nor even most, were committed to the actual operations. Some sources suggest that the IAF combat aircraft were deployed in the following numbers:

Srinagar – 34 (MiG-21, MiG-23BN, MiG-27)
Awantipur – 28 (MiG-21, MiG-29, Jaguar)
Udhampur – 12 (MiG-21)
Pathankot – 30 (MiG-21, MiG-23BN)
Adampur – 46 (Mirage 2000, MiG-29, Jaguar)[27]

Losses

Early on, the IAF suffered some losses. On 21 May 1999, an IAF Canberra PR57 from 106 Squadron on a strategic reconnaissance mission, flown by Wing Commander C.H. Kulkarni, Squadron Leader A. Perumal and Squadron Leader U.K. Jha, was hit by a Chinese-made Anza infrared surface-to-air missile. The plane was able to return, albeit on one engine, to the IAF base at Srinagar, with no loss of life or serious injury and the aircraft was eventually repaired.[28]

The IAF suffered two unfortunate casualties over the Batalik sector while striking against the major Pakistani logistics hub at Muntho Dalo. On 27 May, a MiG-27 of No. 9 Squadron crashed due to an engine flame-out apparently caused by the ingestion of gas from a strafing run using its main gun. The pilot, Flight Lieutenant K. Nachiketa Rao, was forced to eject and was taken prisoner. A MiG-21M, piloted by Squadron Leader Ajay Ahuja of No. 17 Squadron, was escorting the MiG-21 and continued to orbit the site trying to provide cover. This MiG-21 was struck by a Stinger MANPAD fired by Captain Waheed of 5 NLI. Squadron Leader Ahuja was thought by the Indian military to have survived the crash but was later killed. Allegedly the body of Ahuja bore two point-blank bullet wounds, indicating a post-crash death.[29]

On 28 May, in the Tololing sector, a Mi-17 was shot down with the loss of all four of the crew when it was hit by a Stinger missile. The Mi-17, which had a malfunctioning flare dispenser, was part of a four-ship flight of 129 HU. These helicopters were using 57mm rockets to engage a fortified position and had successfully dodged several Stingers before the third ship of the formation was hit after evading three missiles on its own.

MiGs and Mirages in the Strike Role[30]

The Indian Air Force flew its first air support missions on 26 May 1999, operating from the Indian airfields of Srinagar, Avantipur and Adampur. The strike aircraft included MiG-21s – mainly MiG-21Ms, MiG-23BNs, MiG-27s with a few sorties flown by Jaguars principally as reconnaissance aircraft and Mil Mi-17s as helicopter gunships. The fixed-wing aircraft were employed principally with 250kg and 500kg bombs but also made extensive use of unguided 57mm rockets to strike at insurgent positions.

The IAF's fleet of MiG-21s, MiG-23BNs and MiG-27s, was neither equipped with modern attack nor navigation systems and only a handful had been fitted with chaff and flare dispensers. Pilots strapped GPS gadgets obtained on the open market to their thighs or held them if a hand was free. Despite the use of 57mm rockets, the IAFs' MiG fleet was also tasked with dropping leftover 1,000lb bombs from the 1965 and 1971 wars on the advice of the Indian Army which selected impact points, sought to trigger avalanches or landslides in order to cut off Pakistani lines of communication and evacuation.

The effectiveness of ordnance at that altitude and in that environment was severely compromised. Even with computerised weapons aiming systems, unguided bombs were significantly less effective in mountainous terrain and there is little doubt that the use of 250kg and 500kg bombs was significantly less effective than they might have been at lower elevations. While near misses could do damage in the plains, in the mountains, however, this was most definitely not the case due to the undulating terrain and the masking effects of the said terrain. Moreover, inaccuracy would be exaggerated as, due to the variation in elevation, there would be a magnification of inaccuracy in the linear dimension.[31]

This was very noticeable to the Indian army which found that the majority of the IAF's strikes from MiG-21s, MiG-23BNs and MiG-27s with 500kg and 250kg bombs were relatively ineffective. In contrast, there was a greater level of efficacy with the UB-32 57mm rocket pods which were widely employed by all of the MiGs involved in strike operations. Yet the destructive power of their small warheads in mountainous terrain led to substantially less damage than might be ideal. The use of 250kg unguided bombs from Mirage 2000s was much more effective and this highlighted the requirement for aircraft with effective and accurate and computerised weapons aiming. The requirement for pinpoint accuracy in mountainous terrain is evident but this very terrain and the climatic conditions therein render that accuracy difficult to achieve.

The Mirage 2000 initially flew strike operations with unguided bombs, the first of which were Spanish-made 250kg bombs. Larger weapons such as old 1,000lb and newer 450kg bombs were soon made available for use following clearance and both types of bomb were delivered using the Mirage 2000's onboard CCRP (Computer Controlled Release Point) sighting technique. One interesting aspect of the Mirage 2000 deployment was the push to equip the aircraft with old ordnance directed by its excellent attack system. One step that was taken very early in the conflict was the deployment of the aircraft with 250kg bombs of Spanish origin which had been procured for the Ajeet attack aircraft – an Indian development of the Folland Gnat. These modest weapons could be carried in large numbers by the Mirage 2000 with up to 12 being a routine payload during the Kargil operations.

The Mirage 2000s of the IAF were equipped with the ATLIS II/ Bombe Guidée Laser (BGL) Arcole 1,000kg bombs combination. This was less than ideal for use in the Kargil sector. The Atlis II is a French daylight and clear-weather laser/electro-optical targeting pod and the IAF had specifically ordered the combination for targeting recessed command and control sites/bunkers and controlling the deep-penetrating and highly expensive Arcole LGB.

However, the ATLIS pod had operating problems at very high altitudes and were therefore deemed totally unsuitable for deployment in Kargil. The IAF procured a number of the very much cheaper US Paveway II laser-guided bomb kits for use with the Israeli Litening laser designator pods (LDPs) which were to become extremely popular in IAF service – to this date. However, defective parts could not be replaced thanks to the post-nuclear-test US arms embargo and thus this required IAF remanufacture of the said parts.[32]

A major success was scored between 16 and 17 June. On 16 June, the main supply depot of the intruders was located at Muntho Dalo in the Batalik sector. On 17 June, a strike package of Mirage 2000s destroyed this supply depot as well as administrative and logistics facilities at Muntho Dhalo, using 450kg bombs with CCRP techniques with devastating results, not just physically but also in terms of adversely affecting the morale of the intruders.[33]

Another large logistics camp at Point 4388 in Mashkoh Nallah was spotted using the Litening pod on 23 June and a strike package comprising two Mirage 2000s with unguided bombs hit the site on 24 June. A large number of casualties were apparently caused by this strike but it was not until 10 July that this target was deemed to be completely destroyed.

The Mirage 2000 force was used extensively against the heavily defended Tiger Hill. This was attacked throughout June with a combination of LGBs and unguided bombs. Of particular interest is that the IAF used only nine LGBs in the entire conflict – eight delivered by Mirages and one by a Jaguar. This was largely due to the increasingly effective delivery of unguided munitions which proved to be quite adequate to the task.[34]

The IAF flew its first LGB strike on 24 June in a two-seater Mirage 2000, flown by then Wing Commander R. Nambiar with then Squadron Leader D.K. Patnaik in the rear seat, in the company of another two-seater as back up and a third carrying no less than the then Chief of Air Staff, Air Chief Marshal A.Y. Tipnis, to observe the strike. A nocturnal LGB mission was flown against Tiger Hill on 28 June 1999 with Squadron Leader Patnaik in the front seat and Wing Commander Nambiar in the rear. This strike was extremely effective and its accuracy was confirmed by the Indian Army. The effect of this particular strike was significant enough for it to be deemed instrumental in paving the way for Indian ground forces to recapture Tiger Hill. It should be noted that all LGBs were delivered by two-seaters, with the rear-seat pilot doubling up as a WSO – this would limit the number of aircraft so equipped to only seven (corresponding to the number of two-seat Mirage 2000s in IAF service at the time).[35]

At the time of the Kargil conflict, only one attack squadron in the IAF was fitted with GPS and none of the MiG-21 squadrons – and probably none of the MiG-27 squadrons – the sole unit being a MiG-23BN squadron which had undergone upgrades. The IAF procured hand-held GPS on the open market and these, with the target coordinates available, enabled the pilots on approach to the target to deliver their ordnance at the determined distance from the target. The IAF believed that if the coordinates were accurate, the bombing accuracy would be reasonably good. The IAF was also aware that accuracy would be greater at lower altitudes of about 500 feet above the target areas. In order to limit the efficacy of hostile air defences, these sorties were flown at night, particularly risky without radar support and with aircraft lacking modern navigation aids. Moreover, the issue of the ordnance used by the MiG-21s, MiG-23BNs and MiG-27s was at least partly resolved by improved weapons delivery techniques.

The introduction of the IAF's Mirage 2000H fleet into daily operations substantially improved the destructive impact and accuracy of IAF strike against point targets even with the use of unguided low-drag bombs such as the 450kg bomb and the old Spanish 250kg ordnance. This, as noted above, was thanks to the aircraft's excellent avionics suite that includes a Continuously Computed Release Point (CCRP) system which improves weapons delivery by compensating for target area wind and enables vastly improved accuracy. All that a pilot or WSO needs to do is to designate his intended aim point through his cockpit head-up display (HUD) and then depress an approval button on the aircraft's control stick. This then allows the computer to release the bomb once all the required delivery parameters are achieved. This approach gives a pilot a high degree of confidence that his weapon will be delivered with adequate accuracy.[36]

The IAF demonstrated additional innovation especially on its MiG-21s, which despite their lack of sophistication and effective onboard navigation and attack avionics, soldiered on in the close air support role. The IAF used stopwatches and GPS receivers in their cockpits to conduct night interdiction bombing even with the MiG-21. This was a major achievement for the IAF as, despite its limitations, the MiG-21 proved to be a useful close air support asset with unguided rockets and its increased efficacy would have been valued.

As noted earlier, the IAF and Indian Army designated impact points for old 1,000lb bombs to be delivered so as to create landslides and avalanches that could wreck intruder supply lines. That the IAF retained ordnance from the 1965 and 1971 wars in its inventory was decidedly puzzling, but the ordnance seems to have worked effectively enough. Given this use of ageing ordnance, it is fortunate that the IAF did not consider using its Hawker Hunters!

Effective Support

The efficacy of the IAF's operations cannot be seen in isolation from the broader objectives of the Kargil campaign. While the effects of the MiG-21, MiG-23BN and MiG-27 sorties were largely confined to tactical success and their initial operations with 250kg and 500kg bombs were far from effective, the Mirage 2000s scored at least five successful laser-guided bomb hits on forward supply dumps sites and posts.

Between the commencement of operations on 26 May and the last day of operations of 12 July 1999, it was evident that the IAF had dramatically improved its bombing accuracy. Even at night, which could have had a deleterious effect on bombing accuracy, the willingness of the IAF to conduct operations around the clock meant that the Pakistani infiltrators could not retain their posts and were compelled to withdraw.

Moreover, effective Indian reconnaissance of Pakistani artillery positions tasked with supporting the infiltrators rendered them vulnerable to Indian counter-battery fire of ever-increasing efficacy while simultaneously making their own counter-battery efforts less than effective. In all, the IAF flew a total of 550 strike sorties against positions held by the infiltrators as well as bunkers and supply depots. These positions, bunkers and depots were located through the efforts of about 150 reconnaissance and communications intelligence sorties flown by a variety of IAF platforms.

Though the PAF never rose to challenge the IAF, the service dedicated some 500 sorties which were flown for air defence and for escorting strike packages and reconnaissance aircraft. In total, IAF fighters and strike aircraft flew 1,730 sorties. In addition, there was a substantial effort on the part of the IAF's helicopter and transport fleet which between them flew close to 6,000 sorties to supply and sustain India's combat forces in the theatre as well as to provide much needed services such as medical and casualty evacuation.

The First Two Decades of the New Millennium

The IAF slowly increased the numbers of Su-30MKI squadrons and consolidated around a force of Su-30MKIs, Mirage 2000s, MiG-29s, Jaguars and MiG-21s. The drawdown of the MiG-23 and MiG-27 squadrons was rapid and the aircraft were phased out from about 2004 when the last MiG-23MF squadron was decommissioned (the squadron later converted to the Jaguar). The loss of the MiG-23 and MiG-27 squadrons was accompanied by a significant fall in the numbers of MiG-21s as the MiG-21FLs were phased out and the numbers of MiG-21M/-MFs in service fell quite dramatically as aircraft were increasingly consolidated into fewer squadrons. This also occurred with the MiG-21bis.

The IAF's efforts to procure a modern combat aircraft from abroad degenerated into a farce with the competition for 126 aircraft being won by Dassault Rafale but negotiations were interminably embroiled in disputes over costs. In the end, 36 aircraft were procured. Simultaneously, the IAF was inexplicably lukewarm to the indigenous Tejas Light Combat Aircraft despite it evidently showing promise during trials. It was not until 2016 that the IAF finally woke up to the need for the aircraft and began its induction.

The IAF did make some efforts to modernise its air defences with the procurement of several squadrons of indigenous Akash SAMs and the induction of three A-50 AEW aircraft, followed by three indigenous Netra AEW platforms.

The helicopter fleet of the IAF was strengthened by the induction of Dhruv light helicopters while the training fleet saw the phasing out of the HPT-32 over safety concerns and their replacement by Pilatus PC-7s. However, the biggest boost came about when the IAF finally obtained advanced jet trainers in the form of the BAE Hawk. This plugged a major gap in the force's flight training programme. The transport fleet saw the induction of more Mi-17 variants and the first US assets in decades with the procurement of C-17 and C-130 transports. The latter two types are now among the most prized assets in the IAF, though they are fewer in number than the older An-32 and Il-76 transport aircraft.

Balakot 2019

The IAF's combat assets were committed to action in a stunning fashion when, in response to a Pakistan-based terror organisation – the Jaish-e-Muhammed (JeM) – carrying out a suicide bombing which killed over 40 soldiers of the CRPF in Kashmir, the IAF struck at a JeM training facility at Balakot, deep inside Pakistan.

On 26 February 2019, at least 12 Mirage 2000s conducted the strike with SPICE penetrator bombs while others carried air-to-air missiles for top cover. In addition, the IAF deployed Su-30MKIs and MiG-29s to decoy Pakistani assets while protecting Indian airspace. All of this was done with command and control provided by indigenous Netra AEW platforms as well as a Phalcon to provide additional support. It should be noted that the intended use of Popeye missiles was aborted. A total of 22 IAF aircraft were involved, the Mirages for the actual strike operation with the Su-30MKIs providing air cover plus decoying the Pakistan air force F-16s that responded to the IAF incursion.[37]

The Balakot raid has been much disputed for the damage or lack thereof with conflicting accounts and disputed satellite imagery – some of the latter showing no damage, with others noting that there was some evidence.[38] However, Italian journalist Francesca Marino, determined that between 130 and 170 JeM terrorists had been killed.[39] She also noted the massive Pakistani cover-up to claim minimal to no damage, thus calling their bluff.[40] Support for Marino's position was found when *India Today*'s special investigative team released the Balakot tapes, noting that there was evidence of Pakistani Army as well as JeM casualties.[41]

The next day however, the Pakistan Air Force staged Operation Swift Retort, where a strike package of 24 Pakistani aircraft – a mix of JF-17s, F-16s and Mirage-III/Vs – staged a brilliant raid across the LoC but were effectively thwarted by eight Indian aircraft, including two MiG-21 Bisons which engaged the F-16s. In the melee, one MiG-21 Bison was shot down after it crossed over to the Pakistani side of the LoC. The strike mission was ineffective, with a number of precision guided munitions falling short of their target. India claimed that a MiG-21 Bison downed an F-16 but this is unconfirmed and is at best a probable kill. A tragic incident saw the loss of a Mi-17 to friendly fire when a SPYDER SAM was fired at an IAF helicopter due to poor communications between air traffic control and the SAM crew.

In the aftermath of the Balakot raid, both India and Pakistan began aggressive surveillance opposition with their respective electronic warfare assets along with UAVs. To counter Pakistani UAVs, the IAF deployed its 116 HU with Rudra-armed helicopters to intercept and destroy these UAVs, with two other UAVs being destroyed – one by a SU-30MKI and another by a SPYDER SAM.

The Balakot raid showed up strengths and weaknesses in the IAF, with the force being innovative and determined but having some deficiencies that the IAF had spent the last two years trying to remedy by upgrades, new assets and new weapons, all the while grappling with falling numbers. Furthermore, the IAF was forced to examine its weapons – in particular, its air-to-air missiles – to check their reliability and performance after shortcomings were observed in combat. This set in-train a substantial overhaul and modernisation programme which is still ongoing.

China's Challenge in 2020

Between April and June 2020, China began a series of incursions along the Line of Actual Control, aimed at altering the status quo. This ultimately led to a clash at Galwan which killed 20 Indian soldiers and several Chinese, with China admitting five but the Russian News Agency, TASS, indicating a Chinese toll of 45 killed.[42]

This led to a massive build-up of forces along the LAC on both sides. The IAF redeployed a large number of assets to the region with new forward bases being activated. The IAF deployed a force of its attack helicopters – new Apaches, Rudras and pre-production

An Israeli-made Spyder SAM. This Spyder TEL is from one of four squadrons belonging to the IAF. (PIB)

Light Combat Helicopters – to the region and moved several SAM squadrons to protect its assets. Moreover, the IAF sought to adopt as alert and aggressive a stance as practicable with combat assets practicing air-to-air and air-to-ground operations.

The IAF has now moved to employ its aircraft out of its airbases in Ladakh – Leh being the highest such operational base. The older aircraft such as MiG-21s, MiG-23s and MiG-27s were found to perform poorly at such altitudes – even more so with the Jaguar. Indeed, outside the Mirage 2000 and MiG-29s, the only other aircraft from the 1980s capable of operating from Leh were the Hawker Hunters of No. 20 Squadron and some MiG-23BNs. This has now completely changed and the IAF is now capable of providing a high level of combat air support and with the activation of new air bases and advanced landing grounds, along with the enhanced performance of combat aircraft, the IAF could provide significant combat support to both Indian Army troops facing China but also the 102 Infantry Brigade confronting Pakistan in Siachen.[43]

The IAF's transport fleet has come into its own. It has transported thousands of troops, tanks, artillery pieces, missile systems and tonnes of stores to sustain the army through the extremely harsh winter. At the time of writing, two years into the confrontation, the tension between the two countries shows no signs of diminishing and the possibility of a shooting war is very real.

Mention should be made of the stellar role of the Indian Army Aviation Corps. Flying an assorted force of Cheetah and Chetak light helicopters, supplemented by large numbers of the HAL Dhruv and HAL Rudra, the AAC has earned a reputation for excellent work at high altitudes. In support of 102 Brigade, at altitudes of 18,000 feet, they have sustained troops, though the payload capacity of the Cheetah helicopter allows for the carriage of only two soldiers and rations. To enable a build-up and proper sustaining of the force, the Indian Air Force uses a relay system whereby medium-lift Mi-17 (earlier Mi-8) helicopters would ferry supplies to lower altitudes and thereafter Cheetahs would move supplies to high-altitude positions.

The load carried by Cheetahs could be as low as 25kg and as such many sorties would be needed to accomplish even modest logistical and transport feats. The arrival of the Dhruv, with its improved performance, more powerful engine and better avionics dramatically improves the ability of the IAF and Army Aviation Corps to conduct operations at these extreme altitudes and in poor weather, though the risk of accidents is still very high.[44] The AAC, alongside the helicopters of the IAF – Dhruvs, Mi-17s, Chinooks and the ageing Cheetahs and Chetaks – provides the lifeline to support Indian troops at distant, high-altitude locations where other modes of transport will not reach or function with the necessary consistency.

The IAF thus entered the 2020s in an unusual state – it is increasing the quality of its assets while grappling with a significant decline in its squadron numbers from its peak. Into this mix must be added increasing tensions with China along the Line of Actual Control (LAC) that took place from 2020 onwards. This has pressed the IAF into service in support of the Indian Army's massively increased presence along the LAC.

2
THE DOCTRINE OF THE INDIAN AIR FORCE

Like any military service, the IAF is but a single entity in a broader national security framework. However, India as a whole has been rather reticent about publishing its national security objectives and frameworks. This has not stopped the armed forces from articulating what they perceive to be the country's national security objectives and threats.

Indian Defence Policy Summarised
The thrust of Indian defence policy evolved around what R.G.C. Thomas calls the 'minimalist' perspective of being able to fight one 'full' war and one 'half' war.[1] In military terms, this means that India was capable of fighting a war on land, sea and in the air against Pakistan while being able to fight a defensive border war with China.

Unlike the situation in the 1980s and 1990s, though perspective has once again altered in the post-2010 period, China was seen as ultimately the most significant threat throughout the 1960s and 1970s, and although the 1965 and 1971 wars with Pakistan proved that the Pakistani forces were still a substantial adversary, India's leadership viewed China's increasing power with growing concern. To this end, despite the substantial growth in India's naval capability and a widening gap in Indian air superiority, India has still confined itself to being able to fight a full-scale offensive war with Pakistan, while deploying sufficient forces to fight only a defensive war against China.

Even the growth in India's nuclear and missile capability, which has now reached the stage where India has a considerable ballistic missile potential and a large stock of fissile material, has to be seen in light of the fact that there has been no perceptible change in India's basic strategy versus China and Pakistan. Nuclear weapons are an extension of this 'one and a half wars' capability. No meaningful power projection doctrine has yet emerged in India.

India has moved on from this position to one where India has the capacity to fight 'one full and three half' wars. This means that India would be able to conduct a full war with Pakistan, conduct a defensive border war with China, combined with a latent nuclear weapons capability and the ability to intervene in neighbouring Indian Ocean islands.[2] This may be considered to be India's current strategic position. India has deployed IRBMs and is moving slowly towards full range ICBMs openly; however, there seems to be no evidence of India shifting its basic threat perceptions in the near future. Therefore, Pakistan, and to a lesser extent China, remain, in the view of the Indian government, the two major military threats faced by India and to this end, substantial conventional forces and a significant nuclear potential are seen to be vital for preserving India's ability to deter an attack from either country.

Armed Forces Doctrines
The basic doctrine of the Indian Air Force was expounded in 2012 when the 'Basic Doctrine of the Indian Air Force' was published. This, however, was modified when the Indian armed forces published their joint doctrine in 2017. Both of these documents expound principles which are cited below in their entirety. The 2017 Joint Doctrine, as published, for the first time defined National Security Objectives and National Military Objectives which arise therefrom:

National Security Objectives
5. National Security Objectives flow from and are designed to safeguard our National Interests. National Security Objectives, like interests, influence our political, military, and economic dimensions. They provide a framework for the formulation of National Security Policy and ensuing Strategies. India's National Security Objectives are:

(a) Maintain a credible deterrent capability to safeguard National Interests.

(b) Ensure defence of national territory, air space, maritime zones including our trade routes and cyber space.

(c) Maintain a secure internal environment to guard against threats to our unity and development.

(d) Expand and strengthen "Constructive Engagement" with other Nations to promote regional, global peace and international stability.

National Security Policy
6. National Security Policy is based on our National Security Objectives and the components of National Power, weighed against the prevailing and assessed future domestic and global environment. It shall entail inherent right of self-defence, possession of deterrence capability, strategic autonomy, self-reliance, cooperation, security and friendly relations with countries.

National Security Strategy
7. Our National Security Strategy (NSS) primarily revolves around safeguarding our Nation from any type of internal and external threats/aggression. In addition, our NSS encompasses preservation and strengthening of India's democratic polity, development process, internal stability and unity in its unique multi-cultural settings. Our NSS also addresses the general well-being of our vast population, the vitality of our economy in context of globalisation and the rapidly advancing technological world. A regional and an international environment of peace and

cooperation will facilitate the safeguarding of our interests. Even though we have no formally articulated National Security Policy and Strategy, it does not imply that they do not exist or are not sufficiently understood. Central to our NSS is to maintain an effective conventional and nuclear deterrent capability.

National Military Objectives

8. National Military Objectives (NMOs) accruing out from National Security requirements are as follows:

(a) Prevent war through strategic and conventional deterrence across the full spectrum of military conflict, to ensure the defence of India, our National Interests and sovereignty.

(b) Prosecute military operations to defend territorial integrity and ensure a favourable end state during war to achieve stated/implied political objective(s).

(c) Provide assistance to ensure Internal Security, when called upon to do so.

(d) Be prepared for contingencies at home and abroad to render Humanitarian Assistance and Disaster Relief (HADR), Aid to Civil Authority and International Peacekeeping, when called upon to do so.

(e) Enable required degree of self-sufficiency in defence equipment and technology through indigenization to achieve desired degree of technological independence by 2035.

Armed Forces Doctrine

9. Armed Forces Doctrine flows from our NMOs. The Armed Forces Doctrine provides a foundation upon which the three Services must operate in synergy. The Armed Forces Doctrine underpins the development of Service specific strategies which must complement the former. [3]

What is particularly noteworthy of the Joint Doctrine is that for the first time it seeks to enunciate India's security threats and break them down into External – further divided into traditional and non-traditional – and Internal Security threats:

Security Threats and Challenges

6. India's security environment is defined by a complex interplay of regional and global imperatives and challenges. It is impacted concurrently by the positive forces of global connectivity like economic and social integration, on one side, and on the other by the obstructive consequences of unpredictability, instability and volatility that connectivity brings.

7. India's pursuit of transformative national growth and development necessitates a peaceful environment across the security spectrum. However, unique threats and challenges related to inherited fault lines adversely affect the security situation of our Nation that are manifesting along within increasingly blurring lines between traditional and non-traditional challenges.

8. External Threats and Challenges.

Our external threats and challenges comprise 'traditional' and 'non-traditional' challenges. They are enumerated below:

(a) Traditional. India's threats primarily emanate from the disputed land borders with our neighbours. Maintaining territorial integrity and preserving National sovereignty continues to remain a major strategic challenge for India. The intensifying competition for natural resources adds an overlay of volatility to existing fault lines and pose challenges that have potential to germinate conflict. Further, transnational threats posed by the activities of State and Non-State sponsored terrorist organisations are exacerbated by the dynamics of intra and inter-State conflicts which pose a danger to regions beyond our primary theatres. India remains concerned about the presence and role of external powers in the IOR, as global geo-politics shifts from the Atlantic Ocean to the Asia-Pacific.

(b) Non-Traditional. The challenges posed by non-traditional security threats range from proxy war to ethnic conflicts, illegal financial flows, small arms transfers, drugs/human trafficking, climate change, environmental disasters, security of energy/resources etc. These challenges are exacerbated by several countries vying to acquire Weapons of Mass Destruction (WMD) and by the competition for natural resources. Their effects on regional stability and the geo-strategic environment are areas of immediate concern. Further, security of our diaspora, resources and establishments abroad, especially in the Middle East / North African regions, which are home to millions of Indians, remain central to our external security paradigm.

9. Internal Threats and Challenges.

Our internal threats manifest in different dimensions, and are briefly described below:

(a) India's multifaceted internal security challenges include an ongoing proxy war in Jammu and Kashmir, insurgencies in some States and organised crime. Left Wing Extremism remains an important challenge that is sapping our National resources, while also impacting the pace of economic development of affected regions. Illegal cross border migrations due to poor socio-economic conditions and/or law and order situations in their home States is another challenge.

(b) Terrorism supported from outside is resulting in the loss of innocent lives. The fragile security environment in the Af-Pak region and neighbouring support to proxy war in Jammu & Kashmir, lends a possibility of it being a conduit for eastward spread of fundamentalist and radical ideologies. Manifestations of these include an engineered radicalised tilt towards such ideology amongst India's youth. Mitigating it requires a multi-faceted approach facilitated by a robust intelligence network.

(c) The easy access to high end technology has increased the threats, making it multi-dimensional. Ensuring sanctity of our land borders, protection of our airspace and long coastline is imperative and remains our priority.

(d) Radicalisation of youth in some States by suspected social media platforms is also a contemporary challenge to National Security. The management of digital environment, which has the ability to manage conflicts through social media, merits high priority in our National Security calculus.[4]

Moreover, the Joint Doctrine has also noted at a shift in India's approach to conflict. India, from perhaps 2016, began a shift to a more aggressive approach and the Joint Doctrine of 2017 noted this as follows:

India in Conflict/War

19. India has moved to a pro-active and pragmatic philosophy to counter various conflict situations. The response to terror provocations could be in the form of 'surgical strikes' and these would be subsumed in the subconventional portion of the spectrum of armed conflict. The possibility of subconventional escalating to a conventional level would be dependent on multiple influences, principally: politically-determined conflict aims;

strategic conjuncture; operational circumstance; international pressures and military readiness. Conflict will be determined or prevented through a process of credible deterrence, coercive diplomacy and conclusively by punitive destruction, disruption and constraint in a nuclear environment across the Spectrum of Conflict. Therefore, undertaking 'Integrated Theatre Battle' with an operationally adaptable force, to ensure decisive victory.[5]

In some ways, the IAF had foreseen some of this shift and in its Basic Doctrine, outlined a much more expansive approach to air operations, redefining its normally understood role.

The Core Doctrine of the Indian Air Force

The 2012 Basic Doctrine of the Indian Air Force speaks to the development of an air strategy that includes a capability to conduct a strategic air campaign alongside a counter-air campaign and a counter-surface force campaign. The latter would involve the use of air power against land and sea-based targets while the former would focus on depleting an enemy's ability to use its own air power against India. The IAF's doctrine notes that the counter air campaign comprises two basic air operations: Offensive Counter Air (OCA) and Defensive Counter Air (DCA), better known in the IAF as Air Defence (AD) – in both cases, it is possible that a combination of missiles and aircraft might be so employed. Similarly, the strategic air campaign consists of conventional and nuclear operations, and the counter surface force campaign consists of air land and maritime air operations. Transport aircraft and most of the IAF's helicopter assets would undertake air mobility operations that would form part of the combat enabling operations, moving supplies and troops as required.[6]

Normally, control of the air should be the first priority for air forces. This permits own air and surface forces to operate more effectively and denies the same to the enemy. The required degree of control is achieved through counter-air operations. Thereafter, the air commander can deliver combat power when and where needed, to attain military objectives at any level of war. This he does by conducting strategic air and counter-surface force campaigns. All the air campaigns can be conducted independently, parallel with, or in support of surface operations.

Levels of Doctrine

The Basic Doctrine of the IAF speaks to the following levels of doctrine, corresponding to the attendant operational levels:

Strategic Level: This doctrine enunciates the fundamental and enduring principles which guide the use of air forces during war and crises. It establishes the framework for the effective use of air power. For example, the tenet that: 'control of air becomes a prerequisite for effectiveness of all military activities' is an enduring principle.

Operational Level: This translates the principles of the basic doctrine into military action by prescribing the proper use of the air forces on the basis of: distinct objectives, force capabilities, broad mission areas and operational environments. An example of an operational doctrine in consonance with strategic doctrine could be: 'AOC-in-C employing his air force in counter air operations by orchestrating a variety of roles to achieve control of the air'.

Tactical Level: This converts basic and operational doctrine by delineating the proper use of specific weapon systems to accomplish detailed objectives. Tactical doctrine prescribes how roles and tasks are to be carried out and is usually published in manuals such as those brought out by the Tactics and Air Combat Development Establishment (TACDE). For example if Mirage 2000 aircraft are flying escort to an airfield attack package, then tactical doctrine would indicate how the Mirage 2000s would be integrated and coordinated within the force package.[7]

With this framework, the IAF went on to define what it saw as its roles and responsibilities in respect of India's military requirements:

AIR CAMPAIGNS AND ENABLING FUNCTIONS

The relevance of the various air campaigns and enabling functions can be better understood by viewing air power employment in its four basic functions. These are:

Control of the Air. This is achieved by a dedicated counter air campaign through offensive counter air and defensive counter air or air defence operations. Their objective is to gain and maintain the required degree of control of air so as to permit effective employment of all facets of air and surface power.

Application of Combat Power. This is achieved through strategic air and counter surface force campaigns. Here, combat power is applied against surface targets and does not include the targets that are specific to the counter air campaign. Typical roles for air land operations are air interdiction, battlefield air interdiction, battlefield air strike, tactical recce, search and strikes while anti shipping and maritime air strikes are undertaken by the maritime air arm.

Enhancing Combat Power. The air power roles contributing to enhancement of combat power increase the mobility, lethality, accuracy, survivability or flexibility of air and surface forces. This is achieved by combat enabling air operations and air mobility operations. Combat enabling air operations include airborne assault, special air operations, special heliborne operations, air-to-air refuelling, electronic warfare, surveillance and reconnaissance, airborne early warning and search and rescue. Combat enabling air operations also involve testing and evaluation and research and development.

Sustaining Combat Power. If air operations are to be successful they need to be sustained and supported by combat enabling ground operations. Runway rehabilitation, CBRN defence, ground defence, passive air defence (including camouflage and concealment), training, administration and HRD are some examples of these.

ROLES

The exact role that an air force will play would depend on the nature of the threat, resources available and the unique nature of the campaign.

Usually, roles envisaged for the air force are as follows:

1. Defence of national and island territories, against attacks from air and space both during peace and war.
2. Deterring an aggressor from carrying out hostile acts and if deterrence fails to mount an effective response.
3. During operations, achieve control of the air to the required degree to provide full freedom of action to the air and surface forces.

4. Applying direct pressure on the enemy's power of resistance by attacking his crucial centres of gravity.
5. Synergising the combat potential of air power with that of the surface forces to achieve joint military aims and objectives.
6. Deploying and employing forces to protect and project the national interests in any out of country contingency operation.
7. Assisting the government in disaster management or humanitarian relief tasks.
8. Executing counter terrorism and counter insurgency operations.
9. Fulfilling international commitments requiring air power assets, consistent with our national policies and interests.
10. Providing viable second-strike capability in case of a nuclear attack.[8]

Nuclear Doctrine and the IAF

The latter point is of additional interest as India has an avowed 'no first use' nuclear doctrine and the IAF clearly sees itself as being part of ensuring a credible and effective second strike in the aftermath of a nuclear attack. Some of the key aspects of this approach were cited in the official Indian nuclear doctrine as follows:

2. India's nuclear doctrine can be summarized as follows:
    i. Building and maintaining a credible minimum deterrent;
    ii. A posture of "No First Use" nuclear weapons will only be used in retaliation against a nuclear attack on Indian territory or on Indian forces anywhere;
    iii. Nuclear retaliation to a first strike will be massive and designed to inflict unacceptable damage.
    iv. Nuclear retaliatory attacks can only be authorised by the civilian political leadership through the Nuclear Command Authority.
    v. Non-use of nuclear weapons against non-nuclear weapon states;
    vi. However, in the event of a major attack against India, or Indian forces anywhere, by biological or chemical weapons, India will retain the option of retaliating with nuclear weapons;
    vii. A continuance of strict controls on export of nuclear and missile related materials and technologies, participation in the Fissile Material Cutoff Treaty negotiations, and continued observance of the moratorium on nuclear tests.
    viii. Continued commitment to the goal of a nuclear weapon free world, through global, verifiable and non-discriminatory nuclear disarmament.[9]

As will be discussed in a later chapter, the IAF was the first of India's armed forces to have a clearly defined nuclear strike role and as such was instrumental in shaping the early incarnations of the Indian nuclear deterrent. Therefore its own doctrine, at least in respect of its nuclear role, has been shaped by India's nuclear doctrine.

Doctrines and Demonstrated Capabilities

If one examines the defined threats, objectives, roles envisaged and the levels of potential operations, even with its reduced force levels, the IAF has sought to position itself to fulfil these roles and to play a greater part in the military part of India's national security.

This has become ever more important as India's tensions with China have increased dramatically since 2020 and the need to sustain forces in the region of Ladakh, as well as to maintain combat capabilities sufficient to conduct operations against Pakistan as well as China, has meant that the IAF has had to tailor its limited forces to meet its critical operational requirements, simultaneously enhancing capabilities.

The IAF, even with depleted force levels, is fully capable of its primary mission of defending Indian airspace against current threats from Pakistan and China. This has been done by improvements in technology, improved surveillance and upgrades to ground-based air defences. This will further be enhanced when an Air Defence Command, integrating all national air defence assets, is created to better enhance and coordinate the air defence assets of the respective services.[10]

In respect of preserving India's interests and conducting punitive strikes, the 2019 Balakot air strikes were a demonstration of a major shift in India's willingness to deploy the IAF as a retaliatory instrument against Pakistan. While the practical effects of the air strikes might have been debated by some – though there is ample evidence to show that they had some success – the fact is that the IAF was used to retaliate for the Pulwama terrorist attack and effectively called the long-standing Pakistani nuclear bluff that had hitherto stopped India from retaliating.

Yet, even in demonstrating this resolve, the strike and subsequent air battle in Kashmir showed up areas of concern for the IAF, particularly their force levels and, more importantly, capabilities. This disconnect between doctrine and the IAF's force levels is being rectified and will be discussed but when examining the IAF in its current state, the connection between its doctrine, its existing force levels and its desired structure must be taken into consideration and examined as the IAF's modernisation programmes are evaluated.

Another lesser discussed aspect of the IAF's doctrine is its limited, but nonetheless important contribution to internal security by the provision of air transport and casualty evacuation for the country's federal police forces during counter-insurgency operations. That the IAF is acknowledging the need for this capability is an important shift in its conventionally conceptualised operational parameters and is more reflective of reality. While the IAF refuses to commit any strike or other combat assets to counter-insurgency operations, it has been compelled to allow its helicopters to fire back in self-defence during anti-Naxalite operations which have seen the loss of one helicopter to terrorist ground-fire.

Out of Country Deployment

The IAF has considerable capabilities to conduct combat operations, not only against countries with which it has land borders, but against overseas targets. However, in respect of its power projection capability, the IAF has demonstrated an ability to intervene at relatively long ranges and to deploy troops overseas.

The IAF's air transport assets and helicopters are used for a variety of HADR duties within the region. In conjunction with the Indian Navy, the IAF has dispatched relief supplies and transported rescue teams within the Asian region. In addition, it has shown itself able to conduct rescue operations in some of the most treacherous terrain and to conduct rescue operations in the worst of weather conditions.

However, one of the Indian Air Force's power-projection capabilities that is inadequately discussed but which highlights the ability of the IAF to deploy forces out of country to preserve India's interests, is the IAF's substantial airlift capability. For an examination of the importance of this capability, we need look no further than the intervention in the Maldives in 1988 – as described in the previous chapter.

3
THE CURRENT STATE OF THE INDIAN AIR FORCE

The Indian intervention, code-named Operation Cactus, remains to date the most dramatic Indian out of area intervention in a crisis and demonstrates the use of air transport as a means of power projection, besides protecting a friendly foreign government. The operation thus represented a doctrinal shift from the IAF as a tactical air force to one which embraced overseas power projection where its capabilities allowed.

Through the Joint Doctrine of the Indian Armed Forces of 2017 a clear statement of India's National Security Objectives, their threat perceptions and the approach India was seeking to take to deal with its security challenges has been put into words. Five years earlier, however, the IAF had through its Basic Doctrine sought to develop its thoughts regarding what air operations were likely to emerge with reference to those very security challenges and it has tried to develop its capabilities, its force levels and to balance its assets to meet its objectives.

The Indian Air Force is the world's fourth largest, with over 12,000 officers and well over 140,000 other ranks, plus thousands of civilian employees. Despite going through a period of decline in respect of the number of combat squadrons, the IAF is still large and highly capable. Looking at the overall strength of the aviation assets of the force, the air defence network and the network of surface-to-air missiles would show that the IAF has maintained a combat capability that is effective and competent to perform its assigned tasks. This is not to say that improvements are not needed. However, appreciating the size, scope and assets available to the IAF, its capability can be seen.

Organisation

The Indian Air Force is organised into five operational commands which are in turn organised into 48 Wings and 19 Forward Base Support Units, the latter having no permanently attached aircraft but which act as transit bases in peacetime and which can be activated as fully-fledged bases in wartime. Each of the IAF's Wings have two to three squadrons attached to them at their respective home bases.

Table 1: Indian Air Force Commands[1]	
Name	Headquarters
Central Air Command (CAC)	Allahabad, Uttar Pradesh
Eastern Air Command (EAC)	Shillong, Meghalaya
Maintenance Command (MC)+	Nagpur, Maharashtra
South Western Air Command (SWAC)	Gandhinagar, Gujarat
Southern Air Command (SAC)	Thiruvananthapuram, Kerala
Training Command (TC)+	Bangalore, Karnataka
Western Air Command (WAC)	New Delhi

Table 2: IAF Wings – Number and Location[2]	
Wing number	Location
1	Srinagar
2	Lohegaon (Pune)
3	Palam (Delhi)
4	Agra
5	Kalaikunda
6	Barrackpore
7	Ambala
8	Adampur
9	Halwara
10	Jorhat
11	Tezpur
12	Chandigarh
14	Chabua
15	Bareilly
16	Hashimara
17	Gorakhpur
18	Pathankot
19	Guwahati
20	Baghdogra
21	Leh
22	Kumbhirgram (Silchar)
23	Jammu
24	Chandigarh
25	Rajokri (Delhi)
26	Thane
27	Bhuj
28	Hindon (Ghaziabad)
29	Bamrauli (Allahabad)
30	Sarasawa (Sahranpur)
31	Agra
32	Jodhpur
33	Jamnagar
34	Bhisiana (Bhatinda)
35	Suratgarh
36	Makarpura (Baroda)
37	Car Nicobar
38	-Unknown-
39	Udhampur
40	Maharajpur (Gwalior)
41	Jaisalmer
42	Mohanbari

43	Sulur (Coimbatore)
44	Nagpur (Sonegaon)
45	Sirsa
46	Nal (Bikaner)
47	Tanjavur
48	Phalodi
49	Naliya

Table 3: Forward Base Support Units[3]	
FBSU number	Location
1	Rajasansi (Amritsar)
2	Sirsa (Now 45W)
3	Nal (Now 46W)
4	Ahmedabad
5	Uttarlai
8	Awantipur
9	Udhampur (Now 39W)
10	Bhisiana (Now 34W)
11	Suratgarh (Now 35W)
12	Naliya (Now 49W)
14	Purnea
15	Port Blair
17	Trivandrum
19	Thoise

Table 4: Air Force Stations			
No.	Name	Location	Type
401	AFS Suryalanka	Bapatla AP	Missile
402	AFS Chakeri	Kanpur	Maintenance
403	AFS Kumbhirgram	Silchar Assam	AFS
404	AFS Begumpet	Secunderabad	Flying Training Establishment
405	AFS Sambre	Belgaum	Non-Flying Training Establishment
406	AFS Bidar	N Karnataka	Flying Training Establishment
407	AFS Coimbatore	Coimbatore	Non-Flying Training Establishment
408	AFS Hakimpet	Secunderabad	Flying Training Establishment
409	AFS Dundigal	Dundigal	Flying Training Establishment
410	AFS Jalahalli	Bangalore	Airframe
411	AFS Cotton Green	Mumbai	Missile
412	AFS New Delhi	Safdarjung	
413	AFS Tambaram	Madras	Flying Training Establishment
414	AFS Yelahanka	Bangalore	Flying Training Establishment
415	AFS Basant Nagar	New Delhi	
416	AFS Vimanapura	Bangalore	Flying Training Establishment
	AFS Chimney Hill	Bangalore	Radar
	AFS Laitkor Peak	Shillong	Radar
	AFS Sambre	Belgaum	Non-Flying Training Establishment
	AFS Salua	Kalaikunda	Radar
	AFS Digaru		

In addition to these operational bases and commands, there are also several Air Force Stations which serve administrative and training purposes.[4]

Combat Assets

The Indian Air Force has an effective strength of 31 combat squadrons. These include 13 squadrons of the Su-30MKI, three each of the MiG-29 and Mirage 2000 (currently undergoing an upgrade), six of the Jaguar (at the initial stage of an upgrade process), two of the Dassault Rafale, two of the Tejas Mk.1 (one under formation) and four of the MiG-21 Bison. In addition, the equivalent of half a squadron with the Tactics and Air Combat Development Establishment (TACDE) operates a variety of attached aircraft – Mirage 2000s, Su-30MKIs and Jaguars in the main.

The last squadrons of MiG-21M, MiG-21bis and MiG-27s have now been phased out of service. It is to be noted that the peak strength of the Indian Air Force was approximately 39.5 combat squadrons, with four MiG-23MF/-BN and six MiG-27ML squadrons forming the core of the strike assets and some seventeen MiG-21 FL/M/MF/bis squadrons forming the bulk of the air defence units. These were, at the time, complemented by the Jaguar, Mirage 2000 and MiG-29 squadrons, which added a high-technology cutting edge to an otherwise mediocre force. Since then, the MiG-21, MiG-23 and MiG-27 squadrons have been completely phased out with some being replaced by Su-30MKIs, and others by Rafales and the Tejas but there remains a gaping void in terms of the IAF's squadron strength that will only partially be filled in the near future.

Combat Assets – Interceptors

India's air defences currently rely on a mix of Su-30MKI, MiG-21/-29 and Mirage 2000 interceptors and 38 squadrons of surface-to-air missiles. These are integrated into an air defence ground environment system and this will be discussed at length in a separate chapter.

Fighter Aircraft

India's air defences are heavily reliant on a force of Su-30MKI interceptors – 13 squadrons of which will form the backbone of the IAF in years to come. Augmenting these are three squadrons each of MiG-29s and Mirage 2000s which are currently undergoing a deep upgrade aimed at enhancing their capabilities – not least of which is their air defence potential. Finally, four squadrons of MiG-21 Bison aircraft complete the Indian interceptor force and these aircraft are set to serve at least until 2024. Upgrades to the Mirage 2000s, MiG-29s and eventually the Su-30MKI fleet will keep these aircraft viable for years to come. This marks a great change from the decades

A front view of a Su-30MKI, 'fully loaded' for display purposes, including (from left to right): R-77 air-to-air missile, a multiple ejector rack with six 250kg bombs, Brahmos supersonic cruise missile (under the centerline), another R-77, and six additional 250kg bombs. (IAF)

Dassault Rafale multi-role fighter-bomber, which marks a quantum leap in capability for the IAF. (IAF)

The Mirage 2000H proved the most popular and most versatile multi-role type in the IAF of the 1990s and 2000s. (IAF)

between 1980 and 2015 where a very large proportion of the Indian interceptor force was comprised of non-upgraded MiG-21s with no capability to fire beyond-visual range missiles. The two Tejas squadrons, which have the capability to fire Derby beyond-visual range missiles adds to this force. The induction of two squadrons of Dassault Rafale combat aircraft with Mica and Meteor air-to-air missiles are the most potent air combat and interception assets in the IAF.

Strike Assets

At its peak, the Indian Air Force operated no fewer than eight MiG-27 squadrons, three MiG-23BN squadrons and five Jaguar squadrons. A sixth Jaguar squadron was formed following the decommissioning of the MiG-23MF interceptors. The MiG-23BN/-27s carried the burden of the close-air support and even the counter-air role, being equipped with a wide variety of unguided munitions and runway denial bombs. In addition, some MiG-27s were equipped to launch Kh-25MP anti-radiation missiles, Kh-23 and Kh-25 air-to-surface missiles and Kh-29 laser-guided missiles. With these eleven MiG-23/-27 squadrons now being reduced to nothing with all aircraft being phased out, there is a large gap in the IAF's strike tactical capability. A deep upgrade of the Jaguar, which includes new engines plus an advanced DARIN III navigation and attack system will keep the Jaguar a viable strike platform for a considerable period of time.

Transport and Helicopter Units[7]

The Indian Air Force operates over 200 transport aircraft and over 300 helicopters. In the case of the latter, the Russian Mil Mi-17 dominates the force, supplemented by Chetak and Cheetah helicopters. In the case of transport aircraft, the An-

A Jaguar/Shamsher armed with a pair of 500lbs bombs installed in tandem under the centreline. The DARIN III upgrade makes the type compatible with a wide range of precision guided ammunition. (IAF)

32 fulfils most transport tasks, augmented by larger types such as the C-130J, the Il-76 and the C-17. New helicopters such as the Chinook and Apache have been inducted.[8]

Transport Aircraft – Unsung Heroes[9]

The Indian Air Force operates a large transport fleet that numbers well over 200 aircraft. These assets can be divided into VIP, light, medium and heavy transport aircraft. This fleet is relatively unique in the IAF in that only its light transports – four squadron equivalents of HS-748 Avros and two squadrons plus one flight of Do-228s are locally made.

VIP transport rests on four each of the Boeing 737 and the Embraer 135BJ jets with the former providing long-range VIP transport and the latter handling shorter-range assignments. It is interesting to note that the IAF's VIP transport fleet is modest in size, smaller than that of the Mexican Air Force's own force of VIP transports.

The bulk of the transport fleet consists of seven squadrons operating over 100 An-32 medium transport aircraft of which a substantial number have undergone a deep upgrade which will be extended to the whole fleet. Forty aircraft were upgraded in Kiev, Ukraine and an additional 45 have been upgraded by No. 1 Base Repair Depot (BRD) of the IAF. The $400 million An-32 project divides the work share into a Total Technical Life Extensions (TTLE) for 40 aircraft at Antonov-certified plants in Ukraine. The contract then provides for the supply of material and transfer of technology for the upgrade of the remaining 64 aircraft at the IAF's No. 1 Base Repair Depot (BRD) in Kanpur. In addition, there is a parallel three-year, $110 million contract with Motor Sich OJSC in Zaporizhia to upgrade the fleet's AI-20 engines. Unfortunately, this project has

A Jaguar/Shamsher seen undergoing maintenance. Notable is the overwing installation of an ASRAAM air-to-air missile. (NAL)

Table 5: IAF Aircraft and Squadron Strength (Approximate Numbers)[5]

Combat Aircraft	Over 650
Helicopters	Over 400
Transport Aircraft	Over 250
Trainers	Over 250
TOTAL	Over 1,500 aircraft

been affected badly by strained ties between Russia and the Ukraine following the annexation of the Crimea.[10]

The HS-748 fleet is extremely old and it is usual for the IAF to continue flying so long past its service life in other air forces. It is currently still used for light transport and liaison purposes but also functions in the cargo transport role as needed, as several of the aircraft have large cargo doors which have proven to be useful over the decades. A replacement of the type by the EADS C-295 transport which is to be manufactured with private sector involvement has not yet seen progress.

The heavy transport fleet, which includes six Il-78 aerial refuelling aircraft, comprises modest quantities of Il-76 (17), C-130 (12) and C-17 (11). The C-130 is the smallest of the three but has significantly more capacity than the An-32 and as such cannot be considered a medium transport. The C-17 was purchased to augment the Il-76 fleet which has been in service for several decades but with C-17 production being halted, the fleet is unlikely to grow any further and replacements of the Il-76 will be many years in the future.

Helicopter Units

The IAF helicopter fleet is large and diverse. The backbone of the fleet is provided by no fewer than 151 Mi-17V5 helicopters with possibly 48 more to be ordered.[11] These augment some 160 Mi-17s and between the two variants of the basic Mi-17, completely dominate the Indian medium helicopter inventory with no signs of either dissatisfaction or replacement plans on the part of the IAF. The Mi-17s have also been armed with machine guns and unguided rockets to give them a useful combat capability as demonstrated in 1999.

Heavy transport capability rests with one unit of three Mi-26 helicopters though their augmentation with 15 Chinook helicopters is in progress with deliveries of the Chinook having already been completed, allowing an upgrade of the Mi-26s to be undertaken.[12]

Light observation and liaison tasks are performed by six units of Chetak and two of Cheetah helicopters and though still delivering excellent service, the two types are in dire need of replacement. It is anticipated that HAL's Light Utility Helicopter and the Kamov Ka-226 will share both the IAF's and Indian Army's requirement for the type. Several helicopter units also operate the HAL Dhruv in the light transport role.[13]

Attack helicopters have never figured prominently in the IAF's fleet and only two helicopter units were equipped with Mi-25 and Mi-35 helicopters respectively. Some of the former have been transferred to the Afghan Air Force. Replacement of the two types by 22 Apache helicopters, 12 being of the Longbow variant, has been completed. A requirement for an additional 65 attack helicopters is likely to be met by HAL's Light Combat Helicopter, two of which have been delivered. Sixteen armed Rudra helicopters – a variant of the Dhruv – are also being delivered with all 16 having been delivered to the IAF by 2021. The armed helicopter fleet of the IAF is thus likely to see a significant improvement in capability as well as an expansion in its fleet of combat assets very soon.[14]

Table 6: IAF Aircraft Fleet Strength [6]			
Type	Known Squadrons	Total No of Squadrons	Total Aircraft
MiG-29	28, 47, 223	3	66
Mirage 2000H	1, 7, 9	3	49
MiG-21 Bison	3, 4, 21, 51	4	90–100
Sukhoi-30 MKI	2, 8, 15, 20, 22, 24, 30, 31, 102, 106, 220, 222	13	262
HAL Tejas	18, 45	1+1 raising	26–30
Dassault Rafale	17 & 101	2	36
Jaguar IS	5, 6, 14, 16, 27, 224	6	114
Total Combat Aircraft			Approximately 650
Mi-17 / Mi-17V5	127, 128, 129, 130, 152, 153, 154, 155, 156, 157, 158, 159	At least 12	223 (151 Mi-17V5)
Mi-24/35	104	1	10
Apache	125	1	12+10 to be delivered
Chinook	126	1	15
Mi-26	126	1	3
Chetak	111, 115, 116, 141SSS+F, 142SSS	4	48
Cheetah	114, 131F, 132F,	2	24
Dhruv	117, 151	4	48
Rudra		1	16
Total Helicopters			Approximately 400
C-17	81	1	11
Il-76 MD	44, 25	2	17
Il-78 MKI	78	1	6
A-50E	50	1	3
Netra	200	1	3
C-130J	77	1	12
An-32 (119)	12, 25 'B', 33, 43, 48, 49, PTS, TTW	7	104
HS-748 (68)	11, 41 'A', 59, 106 'A', 6 Cmd Flts (2 A/c)	4	64
Do-228 (24)	41, TTW, 6 Comm Flts (2 A/c)	2.5	40
Boeing 737	Air HQ Flt	1	4
Embraer 135BJ	Air HQ Flt	1	4
Total Transports			Over 250
HJT-16 Kiran	AFA(A), AFA(B), FIS, FTW, 52		84
BAe Hawk Mk 132	HOTS, OCU		110
Pilatus PC-7	AFA		75
Total Trainers			Over 250

The IAF acquired 22 AH-64 Apache attack helicopters. (IAF)

The basic Dhruv, sometimes fitted with an EO turret, serves in the transport role in the IAF. (IAF)

The need for a R-77 replacement has been felt with the Astra being designed to fill that role. Already achieving intercepts at a range of 75–100km, the Astra has the potential for much evolution and growth and its initial versions have entered service to supplement and later supplant the R-77 and R-27 missile in IAF service and also to eventually augment the Derby and Mica missiles used by non-Russian aircraft in IAF service – such as the Tejas, Mirage 2000 and Rafale.

The Tejas Mk1 and Mk.1A are currently cleared to fire the Derby BVR missile and the R-73 WVR missile. There are plans to integrate the Israeli Python-V WVR missile and in 2021 this was successfully completed. There are reports that the IAF will try to integrate the AIM-132 ASRAAM air-to-air missile onto the Su-30MKI and this missile has already been ordered to replace the R.550s formerly used by India's Jaguars; the ASRAAM has been evaluated aboard the IAF's Hawk Mk.132 armed trainers.

The Dassault Rafale promises to revolutionise the IAF's air combat capability. In addition to a substantial stock of Mica air-to-air missiles, India has ordered the MBDA Meteor long-range missile which can engage targets at ranges beyond 150km. This promises to completely transform India's air combat capabilities and to confer onto India a significant advantage over its rivals in potential theatres of border conflict.

MUNITIONS – ADDING CAPABILITY

Air-to-Air Missiles[15]

The Mirage 2000 upgrade replaced the Super 530D and R.550 with advanced, longer-ranged Mica air-to-air missiles for both the beyond-visual-range (BVR) and within-visual-range interception (WVR) tasks. The MiG-29s and Su-30MKIs as well as the MiG-21 Bisons now use a mix of R-77 and R-73 missiles for BVR and WVR interception respectively. In addition, the Su-30MKI fleet uses long-range variants of the R-27 missile which can engage targets over 130km away.

Air-to-Surface Ordnance[16]

The IAF operates a wide variety of unguided ordnance with S-24 rockets and 68mm rockets being in fairly widespread service. Indigenously-manufactured High-Speed Low Drag bombs in the 250kg and 450kg range have been deployed extensively and are deployed on almost all Indian aircraft.

Laser-guided bombs, using Israeli Griffin adaptors as well as French Matra 1,000kg weapons, complement Paveway systems in IAF service. India has attempted to make a laser-guided bomb in the form of the Sudarshan with a range of 9km. It has also initiated

The Boeing C-17A is the IAF's most potent transport asset. (IAF)

HAL's Light Combat Helicopter is now entering service. (HAL)

The Rudra is an armed version of the Dhruv helicopter. 116HU employs the type. (IAF)

To augment the stock of Hellfire, AT-6 and Shturm Ataka anti-tank missiles, the Dhruvastra with a 7km range has been cleared for use from the Rudra attack helicopter. Furthermore, development of a stand-off anti-tank missile (SANT) with a range of 15–20km has begun with captive trials being conducted from a Mi-35 helicopter. Finally, in 2018, tests of a 100km range Smart Anti-Airfield Weapon (SAAW) were successfully conducted from a Jaguar aircraft using live warheads. This system is now nearing the end of its development and may be inducted soon. Six SAAWs can be carried by each Jaguar. The type has also been cleared for use from the Hawk Mk.132 to augment its existing armament of unguided rockets and bombs.

The Brahmos missile has been adapted for launch from the Su-30MKI initially as an anti-ship weapon and thereafter in its land-attack version – each with a range of 300–450km. The anti-ship air-launched version has already been successfully tested from the Su-30MKI. Lighter versions are being considered for employment from the Tejas and an air-to-air version of the missile is being contemplated.

Alongside these indigenous systems, India has acquired Harpoon anti-ship missiles for its Jaguar aircraft which are undergoing upgrades, with Kh-59 and Kh-35 missiles being obtained for the upgraded MiG-29s. Furthermore, the Su-30MKI has been observed with Kh-29, Kh-31 and Kh-59 missiles in various versions while TV-guided bombs such as the KAB-500KR are in widespread service.

India's Jaguar and Mirage 2000s also deploy the SPICE-1000 bomb and the Popeye/Crystal Maze air-to-surface missile and the Belouga sub-munition dispenser. It is of interest that while the Jaguar is being upgraded and is capable of delivering laser-guided munitions, it remains to be seen if other stand-off munitions will be integrated. One such option, in particular for use against armoured vehicles, is the Brimstone missile which is on offer to the IAF but has not yet been selected.

a project to manufacture a 50km range next-generation laser-guided bomb.

Furthermore, in 2016, successful tests from a Su-30MKI were carried out in respect of a non-winged glide bomb called the Garuda with a range of 30km and a winged glide bomb called the Garuthmaa with a range of 100km. Testing has commenced for a next-generation anti-radiation missile called the Rudram with a range of 100km to 125km.

Jaguars deployed for maritime-strike purposes are nowadays armed with the US-made AGM-84 Harpoon anti-ship missile, one of which is visible under the centreline of this example. (HAL)

A Spice guided bomb installed under the centreline of a Mirage 2000H. (IAF)

A Rafale of the IAF armed with SCALP and Mica missiles, and releasing decoy flares. (IAF)

The Dassault Rafale as the supplied ordnance will include provision for the carriage of Indian ordnance, including those fitted with laser-guidance kits, the SPICE-1000 bomb and the ALARM anti-radiation missile. The biggest system included with the Rafale is the SCALP stand-off munition which has a range of over 500km and which has the promise of turning the Rafale into a most formidable strike platform.

UAVs – A Growing Force

The IAF operates five squadrons of UAVs plus a Technical Flight. The IAF operates well over 200

Table 7: Known UAV Squadrons[19]					
No.	Unit	Nickname	Raised	Based at	Location
3001	Sqn		27 Nov 2000	34 Wing	Bhisiana (Bhatinda)
3002	Sqn	Cat's Eyes		41 Wing	Jaisalmer
3003	Sqn	Trinetra		23 Wing	Jammu
3004	Sqn		20 Sept 2004	12 FBSU	Naliya
3005	Sqn			8 FBSU	Awantipur
UAV	Tech Flight			34 Wing	Bhisiana (Bhatinda)

Harpy and HAROP loitering munitions, but the core of its UAV fleet is comprised of Searcher and Heron-1 UAVs which are extensively used for surveillance operations. The Indian Army and Navy have separate, and in the case of the Indian Army, much larger, fleets of UAVs of both the Searcher and Heron types for their own, extremely demanding, surveillance purposes.

The total force numbers well over 180 Searcher and Heron UAVs – with over 100 Searchers and over 50 Herons. Project Cheetah seeks to arm at least some of this number with missiles and turn them into UCAVs.[17] In addition, in 2018, the IAF was allowed to purchase 10 Heron-TP missile-armed drones from Israel.[18] Known UAV squadrons in the IAF are as shown in Table 7.

Electronic Warfare and Airborne Early Warning

The Indian Air Force, in conjunction with the Aviation Research Centre (ARC) of the Research and Analysis Wing (RAW), also operates a number of ELINT and SIGINT platforms. These include one Boeing 707 for surveillance and two Global 5000s, three Gulfstream G100 for ELINT and surveillance tasks along with an additional three Gulfstream-III aircraft used for electronic warfare.

The IAF has six Il-78 tankers, and the basic Il-76 airframe has been used to support three A-50 AEW aircraft which use the Israeli Phalcon radar system. Additional AEW support is provided by three Netra AEW systems which use an indigenous radar mounted atop an Embraer EMB-145 platform. The Netra system proved very effective so six of an upgraded version, using an Airbus 319 or 320 platform, have been ordered. The IAF's overall EW and AEW assets are, while very capable, limited in number and thus deployed in a sparing manner, where needed.

Missile Units

While the Indian nuclear weapons delivery systems are controlled by the Strategic Forces Command, the IAF also operates a number of land-based, conventionally armed, surface-to-surface missiles. The IAF operates at least 25 launchers for the Prithvi SS-250 SRBM and also operates at least one squadron of Brahmos land-attack cruise missiles on mobile launchers. These systems are apart from those operated by the Indian Army in the same role. The IAF also operates over 30 squadrons of surface-to-air missiles as part of its air defence network but these will be detailed in a later chapter.

The Indian Air Force – Training Programme and Assets

The Indian Air Force operates a robust and effective flight training system which provides a solid foundation for air combat. At the heart of this is a system that combines aircraft, simulators and a operational conversion system that emphasises building air combat skills before a fighter pilot is declared operational on the various types of aircraft. The Indian Air Force operates a variety of trainer aircraft. Basic training is conducted on the Pilatus PC-7-II while basic and intermediate jet training is conducted on the HJT-16 Kiran trainer. Advanced training and armament training is conducted on the Bae Hawk.[20]

Flight Training in the Indian Air Force

The IAF operates a three-stage flight training programme which involves the use of basic turboprop trainers, an intermediate state with basic jet trainers and an advanced stage with advanced jet trainers. India's fighter pilots begin their flight and academic at the Indian Air Force Academy in Dundigal. This training is comprised of three stages. The first stage is six months of flight-related training which takes place alongside the standard joint services training that all three services receive.

Prior to flying, recruits to the flying branch first receive technical training on a trainer aircraft, and this includes the theoretical teaching of air combat principles plus practical demonstrations of aircraft systems. It should be noted that the IAF accepts trainees who do not come through the National Defence Academy and such trainees must complete a six-month pre-flying training programme before beginning the first stage of training at the IAF Academy.[21]

The progression of flying training in the Indian Air Force has three stages – Basic, Intermediate and Advanced.[22]

Basic Stage: This stage is where new pilots receive simple instruction in flying and build psychomotor skills. This stage uses turboprop aircraft, and formerly used piston-engine aircraft. These aircraft are inherently easy to fly and to maintain and provide *ab initio* training in the safest and most economical way possible. Upon completion of the 55 to 50 hours of flying training, the trainees have acquired basic flying skills and learned to manage a flying machine and undertake basic manoeuvers at relatively low speed.

Intermediate Stage: Compared to the global trend of using turboprop trainers for both basic and intermediate training, with a single 120-hour process, the IAF incorporates 60 hours of intermediate training on simple jet aircraft over 24 weeks and which should consolidate basic skills and teach simple tactical manoeuvring.

Advanced Stage: The advanced stage of flight training brings the trainee pilot to the point where he or she is ready to join a combat squadron and be prepared to fly operational missions. At this stage, complex manoeuvring is taught in offensive and defensive modes and the pilot's skill and competency is honed to a sharp point.

Conversion Training: At a final stage, at squadron level, all IAF combat aircraft have two-seat training versions of the specific types. In the past, aircraft like the MiG-21UM and the MiG-23UB had little combat capability. This has now changed with the training versions of the Mirage 2000 and Rafale having the same combat capability of their single-seat counterparts, and so can take their place alongside these aircraft for combat missions. Even the previously limited MiG-29UB is being upgraded to enhance its combat capability, though not with BVR missiles, and the Jaguar trainer always had capabilities similar to the single-seat version, albeit marginally lower.

Aircraft Used for Training

The Indian Air Force operates a variety of trainer aircraft. Basic training is conducted on the Pilatus PC-7-II while basic and intermediate jet training is conducted on the HJT-16 Kiran trainer. Advanced training and armament training is conducted on the Bae Hawk. The Hawk and PC-7-II are advanced platforms fully incorporating a new training curriculum, cockpit layout and new debrief systems. Compared to the earlier HPT-32, and the ad hoc trainers of the Hawker Hunter and MiG-21FL (which were used as part of HOFTU and MOFTU until the Hawks arrived) the terminology and methods of the IAF have undergone a revolution.

It should be noted that while the IAF pushes to move to a two-stage training process on the PC-7 and Hawk, it faces the problem of corruption issues arising from the procurement of the Pilatus aircraft. Into this space and to meet the IAF's requirement for additional basic turboprop trainers, HAL has developed the HTT-40 trainer which has completed six-turn spin trials and has achieved certification for IAF use with up to 108 being considered. An intermediate jet trainer – the HJT-36 – had issues with spin trials but has resumed testing with extensive modifications. This aircraft may not be ordered by the IAF as it seeks a two-stage training programme, but it is a worthy project.

The IAF uses 75 PC-7 Mk.IIs for the basic training role. (Pilatus)

The Hawk Mk.132 is used both as a trainer and light strike aircraft with rockets, bombs and cannon. (IAF)

Use of Simulators

The Indian Air Force began using simulators for its Hawker Hunter, Su-7 and MiG-21 aircraft but these were relatively simple compared to modern systems. Simulators became available for the An-32 and Do-228 transport aircraft. These systems can hardly be called full mission or full motion simulators but were effectively training aids to introduce pilots to the systems and operational system before allowing pilots to fly the respective aircraft.[23]

The HAL Kiran HJT-16 has a system of simulators comprising a Pilot Procedure Platform (PPP) and a Cockpit Procedure Trainer (CPT), while the Hawk AJT has a much more advanced system but this is still not a full motion simulator though it offers significant capability in regards of introducing trainee pilots to the respective aircraft types. The Pilatus PC-7 also has a simulator.[24]

The first Indian aircraft with a full-motion simulator is the C-130J and this represents a quantum jump in technology and vastly improves the training experience and outcomes for the pilots.[25]

Not to be left behind, the IAF has been acquiring advanced simulators for the MiG-29, Su-30MKI and Mirage 2000 aircraft and might well be expected to acquire the same for its Rafale aircraft having been exposed to these during the pilot conversion programme in France. Five full mission simulators have been contracted for the IAF Su-30MKI fleet, representing an important first step and with full-mission simulators being installed at the Halwara and Ambala air force bases, it can be expected that simulators will play a critical role in the future of IAF air combat training.[26]

ADDITIONAL TRAINING AIDS[27]

ACMI

Air Combat Manoeuvring Instrumentation is the next step to enhancing air combat skill. The IAF has imported many such systems from Israel and these have been observed on several IAF combat types. ACMI systems have four components:

Control and Computation

The Control and Computation Subsystem (CCS) is usually a personal computer mounted on a rack and which runs applications that calculate Time-Space-Position-Information (TSPI).

An indigenous Tejas light combat aircraft, seen while refueling from an Il-78MKI tanker. (IAF)

Transmission Instrumentation
The Transmission Instrumentation Subsystem (TIS) is firmware which is located near a communications tower.

Airborne Instrumentation Subsystem
A GPS unit is installed in each Airborne Instrumentation Subsystem (AIS) pod to calculate its own position, giving the CCS a complete TSPI picture.

Individual Combat Aircrew Display
An ICADS is the display software which receives data from the CCS and which then, on a three-dimensional graphical user interface, displays the data it receives.

The IAF thus follows a multi-stage approach to enhancing the air combat skills of its pilots. From a strong flying foundation on fixed wing aircraft to increasing use of simulators and ACMI systems, the IAF is moving progressively forward to new and innovative ways to improve air combat skills. However, as the sophistication of simulators keeps improving, the IAF needs to continuously seek out the best and most effective technology or develop it locally. The development of a simulator for the Tejas combat aircraft and the Dhruv helicopter are steps in the right direction.[28]

The Garud Commandos
The IAF also operates its own Special Forces Unit. The Garuds are an elite IAF Commando unit which was unveiled in February 2004. The force primarily protects Indian Air Force installations from Fidayeen attacks.[29] The Garuds augment the IAF police and Defence Security Corps which provide perimeter security and are responsible for base protection in peacetime alongside armed airmen, providing specialised intervention capabilities.

Besides providing a base protection force to protect airfields and key assets in hostile environments, some advanced Garud units are trained like the Indian Army's Para SF units and the Naval MARCOS to undertake missions deep behind enemy lines. Operations envisaged during hostilities include: undertaking combat search and rescue; rescue of downed airmen and other forces from behind enemy lines; suppression of enemy air defence (SEAD); destruction of radar and air defence assets; combat control, missile and munitions guidance (using laser designators); and other missions in support of air operations. Apart from protecting air bases from sabotage and attacks by commando raids, they are also tasked to seal off weapons systems, fighter hangars and other major systems during intrusions and conflicts.[30]

The force has a total present strength of approximately 1,080 airmen with plans to increase the force to 1,780.[31] The Garuds are equipped with much the same weapons as India's other Special Forces, which include Tavor rifles, Galatz sniper rifles, Negev light machine guns plus a plethora of locally made equipment. They use locally made Light Bullet Proof Vehicles (LBPVs) made by Ashok Leyland. The LBPV is a version of Lockheed Martin's (LM) CVNG (Common Vehicle Next Gen). These vehicles will be used to move Garud units within air bases as well as provide protection to movements of air force personnel and equipment.

4
MODERNIZING THE IAF'S FLEET – INDUCTIONS AND UPGRADES

The IAF desires a strength of some 42 combat squadrons by the period 2027–32 to meet the contingencies of a two-front war. However, four squadrons of MiG-21 Bison will be phased out by 2025. If no new aircraft are ordered, it is possible that the IAF would be left with 29 combat squadrons by 2025 – an overall deficiency of 13 squadrons when set against its desired strength. Subsequently, one Jaguar squadron is due to be retired by 2027, which would mean an overall deficiency of 14 squadrons by 2027.

Reality versus Desired Capability

Although making up this shortfall by 2027 poses significant challenges, the IAF is not without options. It had planned to acquire an additional five squadrons of Rafales and undoubtedly would still like to do so if permitted. To compensate for this shortfall and to cater for future replacements for aircraft such as the Jaguar and eventually the MiG-29 and Mirage 2000, India has two active plans to bolster force levels. One of these plans involves the procurement of new single and twin-engine fighters, with the latter taking priority. The other involves the procurement of four squadrons of the Tejas Mk.1A variant.

Indigenous Fighter Project: Tejas – Mk.1, Mk.1A and Mk.2

As of 2022, Tejas Mk.1 was serving with No. 45 Squadron of the IAF while No.18 Squadron was well on its way to reaching its full strength with a slightly upgraded version, which includes an inflight refuelling probe. The Tejas Mk.1 has demonstrated the capability to deliver laser-guided bombs, and unguided bombs with good levels of accuracy, and is armed with the Derby, R-73 and the Python-V air-to-air missiles. By 2022, some 32 Tejas fighters have been delivered, in addition to the 16 pre-production aircraft and development prototypes. In addition, the IAF has signalled its intention to enhance the combat capability of the aircraft using the Astra air-to-air missile and also a wider range of guided air-to-ground munitions.

In January 2021, the IAF placed an order for 73 improved Tejas Mk.1A single-seat fighters and 10 two-seat trainers for a price of over $6.2 billion, including a comprehensive training and maintenance infrastructure. The Tejas Mk.1A will be equipped with an Active Electronically Scanned Array (AESA) radar and electronic warfare systems currently missing from the FOC Tejas Mk.1. The Tejas Mk.1A may be the ultimate development of the basic Tejas airframe given its lack of internal volume without necessitating major redesign. Other improvements will include a self-protection jammer and a software defined radio – the latter being retrofitted to all IAF aircraft.[1] The choice of radar is the Elta EL/M-2052 but trials of an Indian radar – the Uttam – have started aboard two pre-production Tejas aircraft – LSP-2 and LSP-3 – with apparently impressive results, raising the possibility that the radar could be integrated into the Tejas Mk.1A.

The Tejas Mk.2 promises to be a much larger machine, with a massively enhanced payload – 3,000kg more than the Tejas Mk.1 with its payload of 3,500kg. The Mk.2 will be powered by the GE414 engine and will feature an AESA radar, most probably a version

A still from a video shown a Tejas releasing an Israeli/South African-made Derby air-to-air missile. (DRDO)

A Tejas test-firing an Israeli-made Python Mk.V air-to-air missile. (DRDO)

of the above mentioned Uttam radar. The aircraft will also have an integrated infrared search and track (IRST) system for passive target acquisition and an indigenous software-defined radio-based tactical data link for secured communication and network-centric warfare capabilities supported by IAF's AFNet digital information grid plus a completely internal, comprehensive self-protection suite. Metal cutting of the Tejas Mk.2 prototype has started with a roll-out expected in August 2022 and a first flight in 2023.[2] Though no contract for the type has been signed, the IAF has already expressed interest in acquiring anywhere between five and ten squadrons of the type.

The AMCA

While it might better be described as an industrial project, there is an active Indian 5th generation fighter project – the Advanced Medium Combat Aircraft (AMCA). The AMCA is planned to become a single-seat, twin-engine, stealth, all-weather swing-role fighter aircraft with supercruise capability. The initial prototype is scheduled to fly in 2025 and the supply of titanium alloy for the construction of the prototype commenced in 2021. The roll-out of the aircraft is expected in 2024 and the prototype will likely fly with two GE414 engines.[3]

Importing New Fighters – New Challenges

In the first quarter of 2018, the Indian Air Force, through the Ministry of Defence, floated a Request for Information (RFI) which began the process to obtain new multi-role combat aircraft to augment its depleting combat strength. The proposal is to procure approximately 110 fighter aircraft (about to be 75 percent single-seat and the rest being two-seat aircraft for conversion training). The procurement should have a maximum of 15 percent aircraft in flyaway state and the remaining 85 percent aircraft will have to be made in India by a Strategic Partner/Indian Production Agency (SP/IPA). Of particular interest is the fact that the two-seat aircraft should retain all operational attributes of the single-seat variant (radar, air-to-air refuelling (AAR) probe, internal gun, weapons and infra-red search and track (IRST) and its workload should be manageable by a single pilot and it should be possible to undertake single-seat operation in all roles.

This project is eerily reminiscent of the ill-fated MMRCA competition for 126 combat aircraft which, despite selecting the Dassault Rafale as the L1 bidder in 2012, proved to be so convoluted that the process had to be scrapped and a deal for the direct import of only 36 aircraft was concluded in 2016. This has left a significant shortfall in the IAF's combat strength and as the procurement of additional Rafales through the direct purchase method seems to be embroiled in a seemingly nonsensical political controversy, the need for additional aircraft is acute. However, the stipulation for local manufacture of 85 percent of the aircraft suggests that the eventual delivery of any aircraft is at least five years away.

A total of seven companies have responded to the RFI to date. The six firms which initially competed in the earlier MMRCA competition are competing once again for this IAF contract worth billions of dollars. The aircraft returning for competition include Boeing's F/A-18E/F Super Hornet, Lockheed Martin's F-16 Fighting Falcon, Dassault Aviation's Rafale, the Eurofighter Typhoon, Saab's Gripen and Russian United Aircraft Corporation's MiG-35. A newcomer, however, is the Sukhoi Aviation Holding Company's Su-35 which has entered the fray for the first time. Overall, there is not a great deal to choose from in respect of flight performance and capabilities of the individual aircraft. This deal will therefore hinge largely on the industrial and technology transfer package that will be required for this project. In a sense the IAF will not be purchasing a fighter aircraft but rather investing in an entirely new ecosystem to build fighter aircraft for the IAF's use.

A MICA-armed Mirage 2000IT seen during take-off. HAL is currently upgrading the entire Mirage fleet to the latest I standard. (HAL)

Old choices as Standbys?

While India grapples with the political firestorm over the acquisition of new fighters, there is a need to replace ageing fighters on a priority basis. To this end, consideration is being given to procuring an additional 12 Su-30MKI fighters from HAL and 21 upgraded MiG-29 fighters from Russia. Though this was proposed in 2019–2020, the formal process only began in January 2021 with Russia responding in July 2021.[4]

Fighter Upgrades

The IAF has been upgrading its legacy aircraft to keep them current and viable for the future. The force of MiG-29s, Jaguars and Mirage 2000s has either completed its upgrade process or has an upgrade programme in progress.

The MiG-29 fleet was upgraded with Russian assistance and will be completed by 2022 with more than two-thirds of the fleet already being upgraded. The upgrade included the replacement of the legacy radar with a new Zhuk-ME airborne radar and compatibility with a wider range of air-to-air and air-to-ground weapons which include R-77 air-to-air and Kh-35, Kh-59, Kh-31 and Kh-29 air-to-surface missiles, turning the aircraft into a multi-role platform.[5] Future plans include the integration of the Astra air-to-air missile and a range of Indian air-to-ground missiles including the Rudram anti-radiation missile.

The Mirage 2000 upgrade is proceeding much more slowly and is only half complete with the project scheduled for completion only in 2023 or 2024. This has been partly due to certification issues with new equipment but is also due to delays caused by the Covid-19 pandemic and its restrictions. The Mirage 2000 upgrade includes a new RDY-3 radar, an upgraded electronic warfare and self-protection suite and compatibility with the Mica air-to-air missile and a wider range of air-to-surface munitions including the Israeli SPICE munition and the Crystal Maze missile.[6]

The upgrade of India's Jaguar fleet will be limited to some 54 to 60 aircraft. The upgrade includes the fitment of an Elta EL/M-2052 AESA radar under a new DARIN III (Display Attack Ranging Inertial Navigation) upgrade, new weapon including the CBU-105 sensor-fused munition, ASRAAM, Harpoon and possibly AGM-88 HARM missiles, and the SAAW anti-airfield system; and electronic warfare and self-protection systems plus a helmet-mounted display system. The rest of the fleet will be overhauled but will be left with the earlier DARIN-II navigation and attack system, lacking a radar.[7]

The first flight of DARIN-III upgraded Jaguar. (HAL)

The SAAW airfield-denial weapon installed on a double-ejector rack on a Jaguar. (HAL)

The IAF's fleet of Su-30MKIs have long been the subject of a potential upgrade programme but this has not started in any meaningful way, though talks with the Russian OEM for an upgrade programme are in progress. It is probable that a new AESA radar and self-protection systems will be fitted but at present, the IAF is enhancing the combat capability of the aircraft by integrating a plethora of new weapons systems on the type. These include the indigenous Astra air-to-air[8] and Rudram anti-radiation missiles.[9] The former will replace the R-77 and R-27 missiles which fared poorly during the air battles of 27 February 2019. However, the IAF has commenced integrating R-37, R-77-1 and R-74 air-to-air missiles to immediately augment the existing legacy inventory and provide an immediate improvement in capability.[10] More recently, it has begun to integrate the Mica air-to-air missile with the Su-30MKI.[11] The Su-30MKI has already been integrated with the Brahmos missile and at least one squadron is now deployed with the type.

A still from a video showing the release of the indigenous Astra medium-range air-to-air missile. (DRDO)

Another test-launch of an Astra – which is planned to completely replace both older Russian air-to-air missile-types presently in service: the R-27 and R-77. (DRDO)

Transports: Breaking HAL's Monopoly

In September 2021, India signed a contract for 56 C-295 transports to replace the HS.748 Avros. While not a huge order, the importance is that 40 of the aircraft will be manufactured by the Indian private sector at Tata Advanced Systems Limited (TASL). This project is the sole transport project that is likely to see fruition in the short term.

Helicopter Fleet – Indigenous Designs Bolster Capability[12]

HAL has invested heavily in helicopter designs and production. For decades, HAL produced the Alouette II as the Cheetah and the Alouette III as the Chetak under licence from France with the Cheetah morphing into the Lancer light attack helicopter. Reengining of the Cheetah produced the Cheetal and the same process on the Chetak produced the Chetan but these only secured orders for some ten aircraft in total. HAL never developed either type into a fully-fledged armed helicopter but did clear the Chetak for SS.11B1 missiles for the IAF and to carry torpedoes for the Indian navy.

The flagship of this indigenous programme is the HAL Dhruv helicopter which serves in all three armed forces, the Indian Coast Guard and the Border Security Force. To date some 300 of these helicopters have been produced, and orders for 76 of a weaponised version called the Rudra have been placed. A new combat helicopter – the Light Combat Helicopter – has begun testing and initial orders for 15 have been placed. In addition, a Light Utility Helicopter to replace the Chetak and Cheetah has also flown with orders of over 100 being anticipated.

The ALH Dhruv

The Advanced Light Helicopter – the Dhruv – is a lightweight utility helicopter with good high altitude performance and was designed and developed by Rotary Wing Research and Design Center (RWR&DC) which is one of the Research and Development sections of HAL.

The Dhruv entered military service in 2002, ten years after its first flight. While the helicopter is designed to meet the requirement of both military and civil operators, there have been few customers for the civil variant.

In its new version the Dhruv Mk.III has improved Shakti-1H engines and is often equipped with a new EW suite. It has superior high-altitude performance and has addressed some issues faced with the earlier variants. It can seat 14 fully equipped troops and is a very survivable design. This version is in production and is increasingly important to the IAF for high-altitude operations.

The HAL Rudra[13]

The HAL Rudra, also known as ALH-WSI (Weapon System Integrated) is an armed version of the HAL Dhruv and may be termed the Dhruv Mk.IV. It was developed by the Rotary Wing Research and Design Center (RWR&DC) and manufactured by HAL just as its unarmed counterpart. The Rudra is not merely an armed helicopter but is equipped with Forward Looking Infrared (FLIR) and Thermal Imaging Sights Interface, a 20mm turret gun, and the capability of carrying 70mm rocket pods, anti-tank guided missiles and air-to-air missiles.

Beginning in September 2011, the Rudra underwent weapons integration trials and trials for its electro-optical systems. These trials included testing the 20mm turret gun and subsequently the firing of both 70mm rockets and MBDA Mistral air-to-air missiles in November 2011. The DHRUVASTRA ATGM has been tested from the Rudra and has been cleared for use.

The Rudra has exceeded the payload and performance requirements at the height of 6km and it has integrated sensors, weapons and electronic warfare suite, plus an upgraded version of the glass cockpit used in the HAL Dhruv Mk-III.

The sensor suite includes stabilised day and night cameras, infrared imaging as well as laser ranging and designation. It also possesses an Integrated Defensive Aids Suite (IDAS) from SAAB, with electronic warfare self-protection which is fully integrated into the glass cockpit. On-board self-defence systems include radar and missile detectors, IR jammer, chaff and flare dispensers.

HAL was awarded with a combined order of 76 Rudras for the Indian Army, the primary customer, and the Indian Air Force. The IAF ordered 16, while the army has ordered 60. These have been delivered and new orders are in the offing.

The Light Combat Helicopter[14]

India's first foray into designing and building a dedicated attack helicopter is the HAL Light Combat Helicopter (LCH). This is a multi-role combat helicopter which was once again designed and developed by Rotary Wing Research and Design Center (RWR&DC) and manufactured by HAL. A small number have been ordered for both the Indian Air Force and the Indian Army with anticipation of much larger orders to follow. The LCH has also significant export potential and once induction is underway into the Indian forces, one may anticipate a more effective export drive.

As might be expected, the design and development of the LCH drew extensively on the earlier Dhruv. By using this rotorcraft as a starting point, HAL has apparently reduced the cost of the programme. On 29 March 2010, the first LCH prototype performed its maiden flight and this was followed by an extensive test programme, involving a total of four prototypes, which was conducted without major incident. These tests were gruelling and during the course of these tests, the LCH gained the distinction of being the first attack helicopter to land in Siachen, following the type having landed at several high-altitude helipads, some of which were as high as 13,600 feet to 15,800 feet. By mid-2016, the LCH was deemed to have completed its performance trials, paving way for the certification of its basic configuration. On 26 August 2017, limited series production of the LCH began with 10 LCH being produced for the army and five for the IAF.

The HAL LCH is a multirole combat helicopter, designed to perform various attack profiles, including at high altitudes where previous Indian attack helicopters such as the Mi-35 and even the new Apache, have suffered from severe performance restrictions. The aircraft has a two-person tandem cockpit to accommodate a pilot and co-pilot/gunner, with the aircraft being developed to perform both close air support and anti-armour missions. In addition, the LCH can perform air defence against slow-moving aerial targets, including both manned aircraft and unmanned aerial vehicles (UAVs), and it is anticipated that the type could be useful for counter-insurgency operations (COIN) (though India has little intention of deploying combat helicopters in that role) plus functioning as an escort to special heliborne operations (SHBO), support of combat search and rescue (CSAR) operations, and armed aerial scouting duties.

The LCH has a narrow fuselage and is equipped with stealth profiling, good armour protection, and is equipped to conduct day and night combat operations with an extensive sensor suite. Self-protection systems inherent to the design include a digital camouflage system, an infrared (IR) suppressor fitted to the engine exhaust, and an exterior covered by canted flat panels to minimise the aircraft's radar cross-section (RCS). Its rotor design lends itself to being extremely agile.

On 17 January 2019, LCH completed weapons trials with the successful firing of a Mistral-2 air-to-air missile at a flying target. As early as 2016, the LCH had successfully completed its weapons trials with both 70mm rockets and the 20mm gun. However, an ATGM is yet to be selected for the type. It is anticipated that the 7km range DHRUVASTRA will eventually be integrated onto the aircraft. Two aircraft have been delivered to the IAF and have been deployed in Ladakh.

The Light Utility Helicopter[15]

The HAL Light Utility Helicopter (LUH) along with its derivative Light Observation Helicopter (LOH) is the long-awaited replacement for the ageing Cheetah and Chetak helicopters licence-produced by HAL and which currently still form a disproportionate share of the Indian helicopter fleet.

LUH is a 3-tonne class highly agile new generation light helicopter which can be employed in multiple roles. The aircraft has a cruise speed of 235km/h, maximum speed of 260km/h, service ceiling of up to 6.5 km, a range of 350 km, a maximum take-off weight of 3.12 tonnes and an empty weight of 1.91 tonnes. The LUH is capable of accommodating two pilots and six passengers, and slung externally, it is capable of carrying cargoes of up to 1 tonne. The aircraft can fulfil a multiplicity of roles including reconnaissance and surveillance (for which new sensor packages will need to be integrated), light transport and liaison and, once inducted, will begin replacing the Cheetah as a major part of the transport assets deployed at Siachen to support the Indian Army.

The LUH is powered by a single 750 KW rated Shakti-1U turboshaft engine derived from Safran Arditen, co-developed by HAL and Turbomeca. The aircraft has a dual channel Full Authority Digital Engine Control (FADEC) system along with a backup fuel control system. The helicopter has a glass cockpit featuring a Smart Cockpit Display System (SCDS) and has skid landing-gear.

Between 2016 and 2018, three prototypes were inducted and the aircraft completed hot weather trials at Nagpur in 2018. Subsequently, testing of the aircraft at sea-level altitudes was completed at Chennai in 2018 and at Puducherry in 2019. In January 2019, the LUH successfully completed cold weather trials. Following that, the LUH undertook successful hot weather trials at high altitudes from August 24 to September 2, 2019. To date, these three prototypes have completed over 550 flights cumulatively as of February 7, 2020. The endurance and reliability of LUH was also successfully tested in hot weather and high-altitude tests and the type should soon be given an initial operational clearance to possibly facilitate its induction into the armed forces. These have a requirement for at least 187 such helicopters with the first 12 being ordered.

The Medium Lift Helicopter[16]

HAL has focused its attention on developing light helicopters with Dhruv, the LCH and LUH being either inducted or at an advanced stage of development. However, the HAL Medium Lift Helicopter (MLH) is a different beast altogether. The MLH is a planned large rotorcraft in the 10-15 tonne class. HAL is currently seeking foreign partners either from Russia or Europe's Airbus Helicopters, with which the company has hopes of developing the MLH and for producing around 350 medium lift helicopters to eventually supplement and thereafter supplant the Mi-17 in Indian service. However, to date, despite a mock-up being displayed, the MLH has not progressed particularly far in its development path though the

IAF has shown some renewed interest in the type, despite its faith the Mi-17 family.

Upgrading the Training Fleet

India has three trainer projects with which it can gain great traction. The HTT-40 project and the HJT-36 intermediate jet trainer are promising designs with the latter being the preferred choice for India's new basic trainer.

In 2019 the Indian HAL HTT-40 turboprop trainer successfully completed six-turn spins to both left and right, thus meeting a crucial requirement for the aircraft to be certified for use by the Indian Air Force.

The HTT-40 is designed to supplement the Pilatus PC-7 Mk.2, 75 of which currently serve as the IAF's basic trainer. The IAF currently uses a force of 75 Pilatus PC-7 trainers, over 100 Kirans and 123 BAe Hawk Mk.132s and is currently short of over 100 training aircraft.

It was planned to increase the PC-7 Mk.2 fleet to some 106 to 181 aircraft. However, a corruption scandal has effectively ended any prospect of further PC-7 orders and while spares will continue to be procured, the HTT-40 may be inducted in significant numbers with an order for 70 with an option for 38 more in the offing.

In contrast, the HJT-36 has had a troubled development and only in 2019 was it resurrected, when on 17 April 2019, HAL resumed flight testing of its Intermediate Jet Trainer (IJT) designated the HJT-36 Sitara.

The Sitara was conceived as a replacement for the HJT-16 Kiran which first flew in 1964 and which continues to provide basic jet training to the Indian Air Force. The HJT-36 first flew in 2003 with the SNECMA Turbomeca Larzac engine of 14.1kN thrust. This was deemed to be underpowered by the IAF and in 2005 a contract was signed to replace it with the NPO Saturn AL-55I with 16.9 kN of thrust.

The aircraft has a maximum speed of Mach 0.8 and a ceiling of 12,000m. With a maximum take-off weight of 4,500kg, the aircraft has five stores-pylons for the armament training role.

While initial testing was promising, the aircraft suffered severe problems during spin recovery tests and was deemed to have significant design defects thus prematurely ending flight testing at that time.

After consultation with British Aerospace Engineering Systems, a redesigning of the tail area was undertaken together with incorporating lessons learned from successful spin recovery trials with its own turboprop trainer, the HTT-40. In 2020, the pivotal spin trials began afresh with much greater success.

The final project is that, in collaboration with BAE, of the Hawk-I combat trainer. The aircraft is equipped with a mission computer in the dual redundant configuration which has additional capabilities such as digital map generation (DMG) to provide improved situational awareness.

The HTT-40 seen during its first flight. (HAL)

The MiG-29UPG includes a dramatic improvement of the type's combat capability. (IAF)

90 YEARS OF THE INDIAN AIR FORCE: PRESENT CAPABILITIES AND FUTURE PROSPECTS

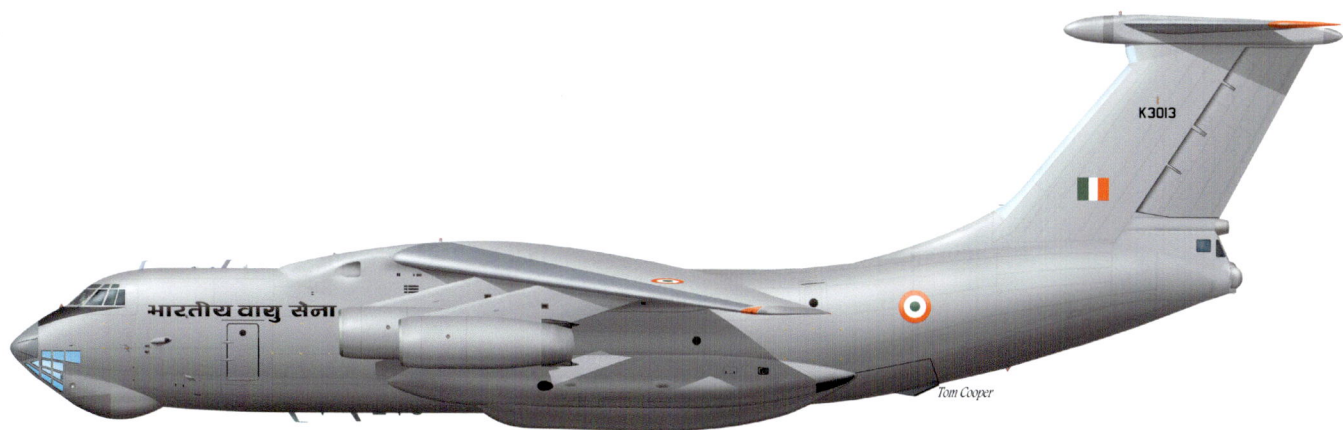

The Ilyushin Il-76 is a multi-purpose transport with strategic reach, powered by four turbofan engines, capable of lifting a payload of 40 tonnes (88,000lbs) over a range of 5,000km (2,700nm). The IAF originally planned to acquire 24 in order to replace the old Antonov An-12s in service with Nos. 25 and 44 Squadrons in the mid-1980s. Only 15 were actually purchased but they proved highly flexible, reliable and capable of operating from short and unprepared runways even in the worst weather conditions. Therefore, the fleet was not only overhauled but, in 2003, reinforced through the addition of two Il-76MDs (serials K2665 and K2901) acquired from Uzbekistan. (Artwork by Tom Cooper)

After years of research and testing on less-capable platforms, in the 1990s India entered cooperation with Israel with the intention of developing an airborne early warning variant of the Il-76. The result was the Beriev A-50EI, an export variant of the A-50 made for the Russian air force but equipped with Viadvigatel PS-90A-76 engines and the Israeli-made EL/W-2090 radar, three of which were manufactured and operated by No. 50 Squadron, IAF. (Artwork by Tom Cooper)

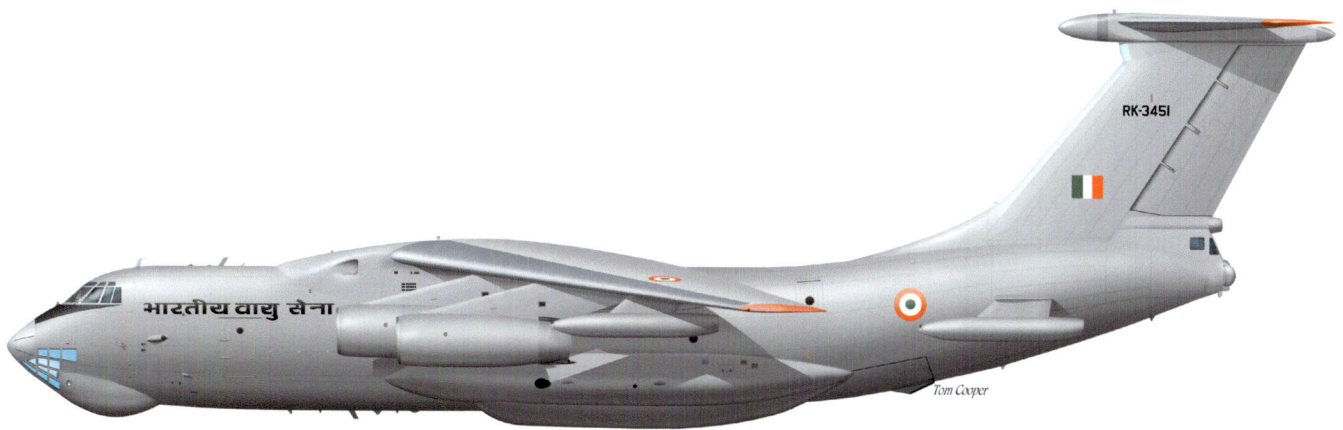

To better support long-range operations of its Su-30MKIs, Jaguars, Mirages, Rafales, and Tejas, in 2003 the IAF purchased two Ilyushin Il-76 transports from TAPO in Tashkent, Uzbekistan. Equipped with French-made avionics and Israeli-made refuelling pods, both were converted to the unique Il-78 MKI tanker configuration, and received serials RK-3451 and RK-3452. Emboldened by this successful experience, the IAF subsequently expanded its tanker fleet to a total of six Il-78MKIs. They are operated by No. 78 Squadron and are presently also used to support A-50EI AWACS and Netra AEW platforms. (Artwork by Tom Cooper)

As a result of a series of studies from the 1990s and early 2000s, in 2003 the IAF and the DRDO launched the development of an AEW aircraft based on the airframe of the Brazilian-made Embraer ERJ.145 light airliner. Equipped with the EL/W-2090 AEW&C system, two of the resulting EMB-145I Netra AWACs entered service in 2012–2013, while one is operated by the DRDO for further development. An order has been placed for three additional Netras, which are to be followed by six of an upgraded version, based on the airframe of the Airbus A.319 or A.320. (Artwork by Tom Cooper)

The IAF originally purchased 36 Dassault Mirage 2000H single-seat interceptors and four Mirage 2000HT two-seat conversion trainers with combat capability in reaction to Pakistan's acquisition of the US-made General Dynamics F-16A/B Fighting Falcon. The Mirages entered service with Nos. 1 and 7 Squadrons, starting in 1986. The fleet was gradually upgraded through the introduction of Atlis II pods and Matra BGL Agricole laser-guided bombs, and then Litening navigation/attack pods from Israel during the 1990s and flew its first combat sorties during the Kargil War of 1999. Ten additional Mirage 2000Hs with improved RDM 7 radars were acquired in 2007, and since 2011 the entire fleet has been undergoing overhauls and upgrades to the Mirage 2000-5 Mk.2 standard – which adds the latest avionics and French-made MICA air-to-air missiles, as illustrated here. (Artwork by Tom Cooper)

Following a lengthy and controversial competition that was to include the acquisition of 126 multi-role combat aircraft, which stalled due to disagreements over licence production in India, in early 2016 New Delhi and Paris eventually signed a contract for the delivery of only 36 aircraft. Deliveries of jets equipped with Meteor air-to-air missiles, synthetic aperture radars, low band frequency jammers, advanced communication systems, radar warning receivers, decoys, and missile approach warning systems began in 2017, and by early 2022 a total of 29 were in operational service with Nos. 17 and 101 Squadrons, IAF. (Artwork by Goran Sudar)

Out of 36 Rafales ordered for the IAF, 28 were single-seat Rafale EH and eight two-seat Rafale DH. While the latter serve as conversion and continuation trainers, they retain full combat capability, and have a highly potent strike role, for which they can be equipped with MBDA SCALP-EG subsonic cruise missiles (shown on the outboard underwing pylon), capable of carrying a nuclear warhead. For self-defence purposes, the type can be armed with Meteor, Mica IR, and Mica EM air-to-air missiles. (Artwork by Goran Sudar)

Following the original contract for 40 Su-30MKIs from 1996, India expanded this to 50 aircraft of this type, and eventually received these in six batches by 2005. Meanwhile, additional contracts were signed for 222 aircraft manufactured under licence by HAL in India, starting in 2004. The IAF thus received a total of 272 Su-30MKIs, including 50 made in Russia and 222 in India, of which 42 were modified to deploy Brahmos cruise missiles. Finally, in 2022, New Delhi placed an order for 12 additional examples to cover attrition. The example shown here, serial SB 024, belonged to the third batch, delivered in 2002 to No. 20 Squadron, with a small squadron patch applied on the fin. (Artwork by Tom Cooper)

The first 28 Su-30MKIs manufactured in India were assembled from parts manufactured in Russia, but by 2010 HAL managed to take over the production of between 50 and 70 percent of all the parts necessary. While prototypes and preproduction aircraft of the Su-30MK/MKI variants were painted in a wide range of camouflage patterns, the definite Su-30MKIs – including Saturn AK-31FP engines with thrust-vectoring – all had the IAF's standard mid-grey overall livery. Radomes and other dielectric parts used to be painted in a slightly lighter grey shade, but many of the HAL examples, and earlier aircraft overhauled by HAL Nasik, have at least their radomes painted in dark grey. (Artwork by Tom Cooper)

Having acquired a total of 843 MiG-21F-13, MiG-21PF, MiG-21FL, MiG-21M/MF, MiG-21bis and MiG-21U/UMs, India was the largest export operator of MiG-21s, and the second largest licence manufacturer of this type. Unsurprisingly, the type formed the backbone of the IAF through the 1970s, 1980s, and 1990s, when negotiations with Israel and Russia were initiated for a possible upgrade of the entire fleet to the latest standard. Eventually, MiG-MAPO won the contract to rework and upgrade 126 MiG-21bis to the MiG-21I-bis standard, generally known as 'Bison', and including the installation of the Kopyo radar and the latest armament, primarily R-77 and R-73 air-to-air missiles. Illustrated is one of the first Bisons at around the time it entered service with No. 3 Squadron in 2002. (Artwork by Tom Cooper)

India acquired a total of 81 MiG-29s between 1986 and 1995, 11 of which were two-seaters (9.51B variant). In 2008, India and Russia agreed an overhaul and upgrade of 64 surviving airframe to the MiG-29UPG standard, including the Zhuk-ME radar, RD-33 Series 3 engines, and a new armament control system compatible with modern weaponry. The work on the first six was undertaken in Russia, while the other jets were reworked at the Ozhar factory starting in 2011. MiG-29UPGs are presently in service with No. 28, 47, and 223 Squadrons, IAF. (Artwork by Tom Cooper)

Searching for a deep strike aircraft with the intention of replacing older English Electric Canberra bombers and HF-24 fighter-bombers, the IAF placed an order for 160 SEPECAT Jaguars in 1979. The order included 40 examples manufactured in the UK; and 120 by HAL. To facilitate rapid service entry, the RAF loaned 18 of its jets and helped with conversion training, and the first Jaguars entered service with the IAF in 1981. Outwardly identical, Indian Jaguars included more powerful engines, compatibility with French-made Matra R.550 Magic Mk. I and Mk. II air-to-air missiles, and the DARIN navigation/attack system. As the aircraft aged, the fleet was put through two upgrades: 125 underwent the DARIN II, starting in 2013, while the latest, DARIN III features an entirely new cockpit and solid state avionics, as well as compatibility with ASRAAM air-to-air missiles and a wide range of precision guided munitions. To improve spares acquisition, in 2018 the IAF purchased 31 disused airframes from France (and two each from UK and Oman), together with a large package of the most critical spares and other items. (Artwork by Tom Cooper)

Emerging from the Light Combat Aircraft (LCA) programme, initiated in the 1980s, the HAL Tejas is a single-engine, multi-role lightweight fighter presently in the process of entering service. The IAF has placed orders for 40 Tejas Mk. 1 and 83 Tejas Mk. 1As, while the development of the Mk. 2 is progressing. This example is illustrated as armed with an Israeli-made LITENING pod and a Russian-made R-73 air-to-air missile. (Artwork by Goran Sudar)

The IAF is planning to acquire a total of 324 Tejas Mk. 1, Mk. 1A, and Mk. 2s. The Mk 2 will be powered by the more powerful General Electric F414 INS6 engine, have added canards and a few other design improvements, enabling an increased payload and additional fuel capacity. In addition to Russian-made weaponry, the first unit to fly Tejas, No. 45 Squadron, already has Israeli-made Python-5 and Israeli/South African-made Derby air-to-air missiles in operational service. (Artwork by Goran Sudar)

After more than 20 years and the most protracted and elaborate procurement processes in the history of India, in 2004 the IAF placed an order for BAE Hawk T.Mk 132 jet trainers. The first 24 were manufactured in the UK, followed by 42 manufactured by HAL between 2008 and 2011. An additional 57 T.132s were ordered in 2008, of which 40 entered service with the IAF, while 17 went to the Indian Naval Aviation. As of early 2022, about 123 Hawks had been delivered and about 110 were still in service, nearly all with the HOTS, for use as basic- and advanced jet trainers, and basic weapons trainers. (Artwork by Tom Cooper)

A Tejas of No. 45 Squadron, seen armed with Israeli/South African-made Derby air-to-air missiles, and equipped with a LITENING navigation/attack pod (underneath the left intake). (Photo by Deb Rana)

A Tejas (serial KH-2017) seen taking off while armed with two Israeli-made Python-5 air-to-air missiles. (Photo by Deb Rana)

No. 126 Squadron, IAF operates 15 Chinook helicopters to augment three much bigger Mil Mi-26s acquired in the 1980s. (IAF)

Armed with cannon, rockets and missiles, the LCH is optimized for high-altitude operations. (Sanjay Simha)

The Light Utility Helicopter will replace the Cheetah and Chetak in service. (Livefist)

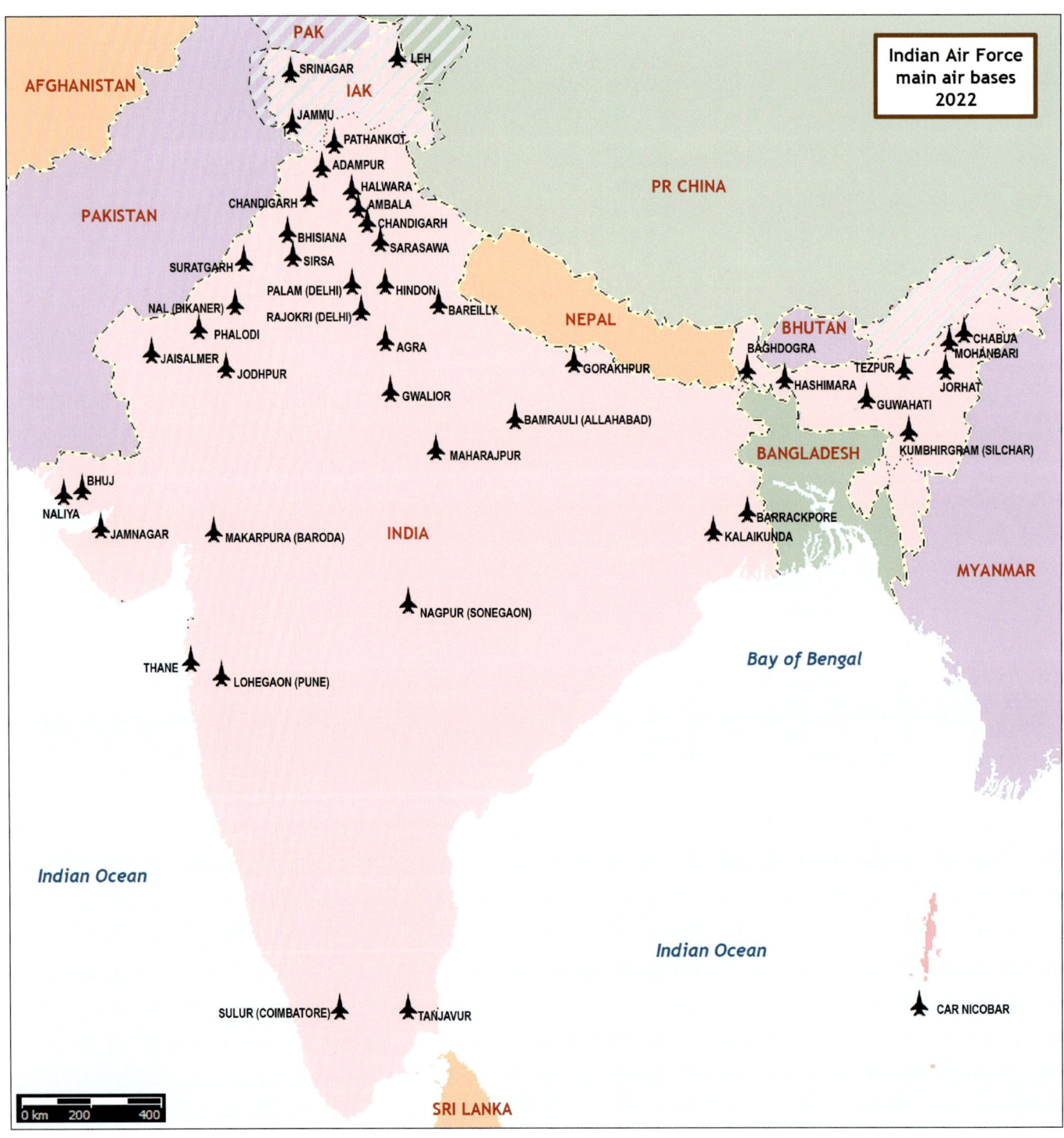

(Map by Goran Sudar and Tom Cooper)

IAK: India-administered Kashmir
PAK: Pakistan-administered Kashmir

The HJT-36 following resumption of spin trials in 2019. (HAL)

Recently the aircraft has also been tested with a variety of ordnance, rockets, bombs and gun-pods being part of its normal inventory while the SAAW anti-airfield weapon and possibly the ASRAAM missile are likely to be integrated.

The IAF has its modernisation challenges ahead. It has reasonably sound plans to upgrade its training inventory and its helicopter plans are well underway. However, its combat fleet is facing the challenge of needing fresh inductions at a time when the force is phasing out legacy aircraft that can no longer be supported. While the IAF has begun to upgrade three of its frontline types, the delay in a deep upgrade of the Su-30MKI is a cause of concern, though the introduction of new weapons does much to overcome some of the shortcomings of the aircraft. Despite the possible purchase of new fighters, the IAF has to place its faith and give its support to India's indigenous fighter designs. The Tejas is maturing into a very good platform and in its Mk.2 version will likely be a good fit to the IAF's requirements.

Hawk-I offers significant upgrade projects for the T.Mk 123 and is undergoing testing, including new armament. (HAL)

5
THE IAF'S GROUND-BASED AIR DEFENCE SYSTEM

India, with its vast airspace, maintains an advanced Air Defence Ground Environment System. This system, along with the civilian Air Traffic Control, is responsible for the detection, identification and, if necessary, the interception of aircraft in Indian airspace. The Air Defence network is also in the process of being upgraded to cater for ballistic missile threats. Before examining the system in detail, a quick overview is in order. India's air defence network is essentially divided into two parts – the Air Defence Ground Environment System (ADGES) and the Base Air Defence Zones (BADZ). These two components are closely linked and share information relating to air defence tasks. The Air Defence Ground Environment System consists of an array of radars along the western and northern borders as well as a network of mobile systems in the north-east and south of the country. Southern India is still peculiarly ill-served by air-defence sensors.

The ADGES network is responsible for overall airspace management and detection of intruders. The ADGES also controls and coordinates the air defences for large area targets. The Base Air Defence Zones, as the name implies, are tasked with the defence of high value targets – air bases, nuclear installations and key military installations. The BADZ is a scaled-down ADGES network, limited to an arc of 100km. The BADZ is a far more concentrated air defence environment than the ADGES and provides the only gap-free air defence cover in most sectors. In addition to these networks, India is now establishing an anti-tactical ballistic missile screen with new radars and weapons. It is not clear whether this will be incorporated into the BADZs or whether it will comprise a separate network. This ATBM screen is slowly taking shape and news of its structure is still awaited. It is noteworthy that India's traditional concern has been its vulnerability to low-level aerial attacks in areas without adequate radar coverage. This is slowly changing to reflect the multiplicity of threats at all altitudes which can emerge.

Indian Air Defences: Sensor Network

The Indian Air Defence Ground Environment System employs a three-tier detection network. While this system is currently in the process of a major modernisation programme, the basic structure of the ADGES network will remain unchanged. The first layer, rather surprisingly, consists of Mobile Observation Posts. These remain among the most reliable of the early-warning mechanisms available to the Indian Air Force. The MOPs consists of two-man teams equipped with a HF/VHF radio set and field glasses. The personnel in the MOP are very well versed in the visual identification of aircraft as well as their general direction of flight. The MOPs are scattered along the borders at random intervals, ranging between 25 and 45 kilometres. The MOPs usually give the first warning of airborne intrusion, the general direction of the attack and, more often than not, the number of aircraft and their type. The MOPs are assisted in this task by personnel from the Indian police forces and Railway Protection Force who are given some training in aircraft identification. These agencies report via a communications system based on both HF/VHF radio sets as well as telephone lines. A more advanced communications system based on fibre optic cables and satellite communications is also available to assist the MOPs in reporting to the radar picket line.

The radar picket line, which lies about 150km behind the MOPs, consists of a number of radar clusters. These comprise three radar stations separated at a distance of the sum of their radii. The equipment issued to these clusters generally comprises one licence-made Soviet ST-68/U and two P-18/-19 radars. These are then flanked by two P-12/-15 radars. The ST-68/U acts as the Control and Reporting Centre (CRC). This may have changed somewhat as the ST-68U, which was plagued with some nagging development problems, has largely replaced older Soviet-made equipment. Moreover, India has been licence-producing the French designed TRS-2215D 3-D surveillance radar for a number of years and has derived an indigenously built radar, PSM-33 Mk.2, from it. These have probably supplanted most of the older Soviet-bloc equipment. It should be pointed out that these radars are all long-range surveillance types with ranges in excess of 300km and offer good performance against targets flying at all altitudes – even those employing electronic countermeasures (ECM). These radar pickets are responsible for giving accurate information on the intruding force to the Air Defence Control Centres (ADCC) located behind the radar picket line. The picket line and the ADCC are separated by a first layer of air defence weapons which are the first to engage the intruders.[1]

The backbone of the Indian Air Defence Ground Environment system is the THD-1955 3-D long-range surveillance radar. This radar, originally of French design, has been licence-produced in India for a number of years. This radar, though somewhat elderly, still has sterling performance characteristics and is capable of maximum detection ranges of up to 1,000km, though in peacetime the Indian Air Force usually limits its power to a 400km detection range. These form the core of the ADCCs. ADCCs also keep in touch with the Base Air Defence Zone (BADZ) control centres.

The BADZ is a scaled-down version of the ADGES configuration and is geared towards the defence of key air bases and other high value targets. The BADZ is limited to an arc of 100km, compared to the hundreds of kilometres in the case of the ADGES system. Like the ADGES, the BADZ consists of three layers. The first of these are the mobile observation posts, followed by a mixed layer of weapons and their associated radars along with a picket line of low-level radars. These are in turn supported by anti-aircraft artillery batteries. This network is controlled by a ST-68U radar. The BADZ provides comprehensive and gap-free coverage over its assigned area of responsibility. Some observers have likened the BADZ set-up to the defence pattern of a carrier battle group. Any aircraft attacking a vital military target, therefore, not only has to get past the ADGES, but also the far more formidable BADZ. This has serious implications for the attacking force.

Upgrading the System

The process of replacing the legacy radars of the Indian Air Force is slowly making progress. A critical element in this is the Arudhra version of the Elta M-2084 radar. Orders for 34 radars of this type were placed in 2009 and between direct supply and the production of an Indian version of the same, this radar will become the backbone of the new air defence network. The Arudhra will act as an early warning and target engagement radar as opposed to the airspace surveillance/air traffic management radars, which currently dominate the IAF's inventory. The IAF will be able to employ the Arudhra-EL/M-2084 to provide early warning of inbound tactical

The IAF operates 34 Arudhra radars – an indigenous variant of the Israeli-made Elta M-2084 system, acquired from 2009. (DRDO)

ballistic missiles and cruise missiles whether air-launched, ground-launched or perhaps even launched from surface or subsurface assets. Installation of these radars will take place in and around Jamnagar, Mumbai and the National Capital Region, to begin with.[2]

Technically speaking, the AESA-based EL/M-2084 represents a quantum leap in technology for the IAF – especially compared to the PSM.33 and TRS-2215D radars that it will replace. It merges all land-based radar functions – weapons location, air surveillance and air defence functions, inclusive of cruise and tactical ballistic missiles. The M-2084 is able to detect and track incoming targets and, in the case of missiles and other such projectiles, it can calculate the anticipated impact and launching points, subsequently passing on target data to the relevant air-defence weapons systems. As an initial step, this system would enable India to detect and potentially engage incoming tactical ballistic missiles but it would not be able to cope with longer range systems.[3]

The sensor network will be completed with a series of low-level and medium-level radars. Low-level surveillance is to be catered for by a total of 67 low-level air transportable radars (LLTR), including nineteen 180km-range, three-dimensional THALES-built Ground Smarter GS-100 radars which were ordered in November 2009. Six of these were supplied directly and the remaining 13 are being licence-assembled by HAL. Each radar will be accompanied by operational and communication shelters, an energy sub-system, a mobility sub-system and personnel living quarters – making them virtually self-contained units. These will augment the 34 active phased-array EL/M-2084 medium-power radars (MPR) detailed above.[4]

In addition, there is a series of Indian-developed systems including the S-band Aslesha three-dimensional micro-radar (developed by the DRDO and built by Bharat Electronics Ltd),

the Army-specific Bharani man-portable radar, and 30 (20 more to be ordered) 180km-range Rohini S-band central acquisition radars. The Aslesha, which weighs 250kg, uses low-probability-of-intercept frequencies to look out for terrain-hugging tactical UAVs over mountainous terrain out to 50km. The IAF has to date ordered 21 of

A Rohini radar system of the IAF, also known as the Central Acquisition Radar. (IAF)

them, and the first deliveries took place in January 2008. In contrast, the Bharani is a two-dimensional L-band gap-filler system now in series-production for the Army. It has a range of 40km and can track up to 100 airborne targets. To date, 16 Bharanis – meant to be used in conjunction with VSHORADS/MANPADS – have been ordered and deliveries are in progress. In addition, also under delivery are 29 THALES Reporter tactical control radars for the army's upgraded ZU-23 and L70/40 air defence guns.[5]

The IAF has a total of three A-50 Phalcon AEW platforms, supplemented by three ERJ-145 based Netra AEW systems developed by DRDO. The IAF is now gearing up to induct new-generation S-band long-range surveillance radars (LRSR) and it is also hoping to acquire an additional nine ELTA Systems-built L-band EL/M-2083 'Airstar' aerostat-mounted high-power radars (HPR) to add to the two already in service, but to date this no progress has been reported in this regard. For the LRSR requirement, a competition is presently underway between the ELTA Systems-built EL/M-2288 AD-STAR, THALES-built Ground Master 400, and SELEX Sistemi Integrati's RAT-31SL.[6] Twenty Ground-Master 400 systems were ordered in 2009 for the IAF.[7] The IAF has also begun to modify some six Airbus A-320 acquired from Air India to accept the Netra's AEW radar and suite while progressing plans for ten more based on the A-330 platform and two more A-50/ Phalcon AEW platforms.

These new radar acquisitions will be integrated with the IAF's existing 32 new mobile control and reporting centres (MCRC), 12 air defence control centres (ADCC), 24 air defence direction centres (ADDC) and some 40 terminal weapons control centres (TWCC) along India's western and north-eastern borders. This will enable the slow but progressive replacement of the above-mentioned existing ST-68U gap-filler radars and related 19ZH6 command-and-control consoles, P-18/NRS-12 and P-19 gap-filler radars. It is anticipated that the LRSR will replace the THD-1955 radars, but this is some way from happening. In the interim, new inductions will enable the replacement of the P-30/NRS-20, P-37 and P-40 gap-filler/target engagement radars, and THALES-built TRS-2215D and BEL-built PSM-33 Mk2 airspace surveillance radars, all of which were inducted in the 1970s and 1980s.[8]

Both the Indian Army and the Indian Navy are desirous of obtaining aerostat-mounted EL M-2083 radars. In the army's case the intention is to acquire six such systems to enable the Corps of Army Air Defence to detect and track both ballistic missiles and terrain-hugging cruise missiles launched from Pakistan, while the Indian Navy is reportedly asking for two EL/M-2083s. Capability-wise, the 1,700kg EL/M-2083 'Airstar' is mounted inside 240-feet-long aerostat that is perched at altitudes of up to 4,000 feet, and uses electronically-steered multi-beam techniques to detect terrain hugging airborne targets – combat aircraft, helicopters, cruise missiles and UAVs – at ranges of up to 300km, while the trajectories of ballistic missiles can be accurately plotted up to 500km away. There has been a renewed emphasis on acquiring these systems on the part of all the three services despite some earlier delays in additional purchases of the type.

These systems would enhance India's TBM defences and cruise missile defences but would offer little by way of assistance in the detection of longer-ranged ballistic missiles which would require radars of considerably greater capability and which have yet to enter service with the Indian armed forces. Yet to see India's BMD efforts as something separate from its overall air defence modernisation programme is to miss the very essence of that programme which seeks to completely transform the Indian air defence network from a 1980s vintage system to the state of the art. This is somewhat different from the approach taken in the United States where the BMD system is distinct from the limited air defence system available for the defence of that country owing to a significantly reduced aerial threat. In contrast, the IAF must cater for manned aircraft, short and long-range cruise missiles as well as tactical and strategic ballistic missiles. The IACCCS, as described below, will be an important component of establishing defences against these threats.

The Integrated Air Command, Control and Communications System

Phase 1 of the Indian Air Force's integrated air command, control and communications system (IACCCS) which seeks to provide a layered, hardened and in-depth air defence command, control and communications network achieved full operational capability by June 2012 with the operationalisation of the AFNET digital information grid which is a fully secure and reliable network owned, operated and managed by the IAF. According to Prasun Sengupta, the IACCC is being established under a two-phase programme at a cost of Rs16,000. It is designed to achieve a robust, survivable network-centric C4I3 infrastructure that will receive direct real-time feeds from existing space-based overhead reconnaissance satellites, ground-based and aerostat-mounted ballistic missile early warning radars and high-altitude-long-endurance unmanned aerial vehicles, and manned airborne early warning and control (AEW&C) platforms. In summary, the IACCCS gives the IAF an automated command and control system for air defence operations as undertaken by the IAF. IACCCS operations will use AFNET enabling the integration of all ground-based and airborne sensors, air defence weapon systems and command and control nodes enabling a coordinated and efficient response to any aerial threat from any sector.[9]

The Indian Air Force has also sought to enhance its airspace management and surveillance capabilities to cater for both peacetime and wartime exigencies. To this end, Sengupta states, the IAF has initiated a multi-phase $1.3 billion programme under which a state-of-the-art joint civil/military sub-continental airspace control system is being developed using the following fundamentals: unity of effort, common procedures, and simplicity. The IAF's terminal area air traffic services and airfield management expertise, and en route airspace/air corridor management are also being upgraded. It is hoped that the final result will be the creation of a vastly expanded air defence identification zone (ADIZ) and provision of a real-time recognised air picture (RAP). The upgraded ADIZ will extend the IAF's airspace management and surveillance coverage (using ground-based sensors) up to 500 nautical miles away from India's territorial boundaries. When fully implemented, new-generation ATCR-33S and SIR-S primary/secondary surveillance radars and their related joint air traffic control and reporting centres (JATCRC) will be operational at IAF air bases in Adampur, Agra, Ambala, Bagdogra, Bareilly, Bhatinda, Bhuj, Bidar, Chabua, Chandigarh, Gorakhpur, Gwalior, Halwara, Hashimara, Hindon, Jaisalmer, Jamnagar, Jodhpur, Jorhat, Kalaikunda, Nal, Naliya, Pathankot, Pune, Sirsa, Suratgarh, Tezpur, Uttarlai, Yelahanka and Zopuitlang in Lunglei district in southern Mizoram.

The IACCCS can be further integrated with the army and navy radar networks and with the civilian airspace surveillance system to enable the provision of an integrated air situation picture for operators to carry out an AD role. It is the intention of the IAF that all of its assets – air, ground and, eventually, space – will be connected to create total situational awareness of a region. For the first time, the IAF will be close to achieving a complete level of integrated coverage of all of India's airspace. This will not only dramatically enhance basic air defence operations but would provide an essential first step on which a BMD detection, tracking and engagement system is to be built. To this end, the IACCCS will also coordinate the early warning and response and interception aspects of India's planned layered, ground-based, two-tier ballistic missile defence (BMD) network that is now is in the development stages and which depends heavily on the IACCCS.

Communications for the IACCCS is provided through the fibre-optic network-based AFNet, which replaces the IAF's troposcatter-based communications network. This system was developed at a cost of Rs10.77 billion in collaboration with US-based Cisco Systems Inc, HCL Infosystems Ltd and Bharat Sanchar Nigam Ltd (BSNL). It incorporates the latest traffic transportation technology in the form of internet protocol (IP) packets over the network using multi-protocol label switching (MPLS).

A large voice-over-internet-protocol (VoIP) layer with stringent quality of service enforcement will facilitate robust, high quality voice, video and conferencing solutions. Having established these critical components, the IACCCS can now be integrated with a large number of new-generation ground-based radars that are in the process of being delivered or are on order. These radars will be able to deal with airspace surveillance in search of airborne targets (such as manned aircraft, ballistic and cruise missiles, attack helicopters and unmanned aerial vehicles), or coastal surveillance or ground surveillance. This would enable India to mount a coordinated response to an attack from any direction and through any medium.[10]

Cruise Missile Defences

The challenge of defending against cruise missiles has assumed as much importance in the Indian context as BMD. While the existing SAMs and radars have some capability against cruise missiles, the deployment of the Chinese ground-launched CJ-10/DF-10A and air-launched K/AKD-20 land-attack cruise missiles or LACM and against the Babur (a DF-10A clone) and Ra'ad LACMs of Pakistan has spurred interest and urgency in the development of an effective cruise missile defence system which would be significantly more effective than the existing network of ageing radars and SAMs.

As for the CMD network's projected deployment sites, two villages in the Alwar and Pali districts of Rajasthan have been selected for the first two QR-SAM Squadrons. Prasun Sengupta claims that Rajasthan's State Forests Department has cleared the acquisition of 850 hectares of land in Khoa in Alwar district, and 350 hectares in Roopnagar, for installing a CMD grid that will protect the western and southern approaches to India's National Capital Region. He further states that site selection work for a similar CMD grid meant for Jamnagar and Mumbai is now in progress.[11]

The Existing SAM Network

The IAF's SAM units comprise 30 squadrons of SA-3b Pechoras and 8 squadrons of SA-8b OSA-AKM systems and are deployed to protect key air bases as well as some major military/industrial centres. Though the SAMs are old, they have been updated periodically and, when operating as part of the BADZ, are deployed in such a manner as to minimise their shortcomings. In addition, a large number of L-40/70 radar directed 40mm anti-aircraft guns and man-portable Igla-1M SAMs are deployed to provide a 'last-ditch' tier of 'hard-kill' defences. It should be pointed out, however, that this system is geared up to the defence of point targets and not for overall area defence. The Igla units are organised into at least 10 flights.

Already, the IAF's SAM units have begun a substantial modernisation process with at least eight of a total of 15 Akash SAM squadrons being inducted to date. Seven squadrons, ordered in 2019, are in the process of delivery and this SAM will significantly enhance the IAF's ground based air defences. This will allow the replacement of roughly half the Pechora squadrons, with eight others being upgraded through overhauls and digitisation. The OSA-AKM squadrons have been augmented by four squadrons of SPYDER SAMs for point defence of vital areas and points.

New SAMs

With this in mind, the Indian Air Force has begun a substantial modernisation of its strategic air defences but while progress has undoubtedly been made, strategic SAM defences remain weak. As noted earlier, the IAF is in the process of inducting eight squadrons of Akash SAMs and has sought to upgrade eight of the Pechora squadrons with an option on four more.[12] In addition, at least nine squadrons of a 70km range MRSAM are being procured with the possibility that several more squadrons may follow which, in conjunction with future Akash procurements, could replace the entire Pechora force.[13] Replacement of the OSA-AKM is in progress with four squadrons of SPYDER SAMs being delivered and a new Quick Reaction SAM under development.

The IAF has not, until recently placed sufficient emphasis on its SAM network and these steps would appear to be the first in a larger process of inducting a new generation of SAMs.

The MRSAM

In 2017, in recognition of the deficiencies in the existing AAD inventory, clearance was given for the procurement of a new India-Israeli developed MRSAM with a 70km range. Initially, some 40 launchers and 200 missiles of this version of the system already in service with the Indian Air Force and the Indian Navy will be procured.

Compared to the existing inventory, the MRSAM will offer the AAD a longer-ranged system but also offers the prospect of enhanced capabilities against various targets. It is capable of shooting down enemy ballistic missiles, aircraft, helicopters, drones, surveillance aircraft and AWACS (Airborne Warning and Control Systems). The missile has a small and lightweight active seeker which dramatically enhances kill probability. The multi-functional phased array radar, based on the naval MSTAR, can detect targets up to a range of 300km and performs the functions of surveillance, multi-target tracking, threat alerts, target assignment and launch of missiles.

The Medium-Range Surface-to-Air Missile (MRSAM) was developed by India's Defence Research and Development Organisation (DRDO) in collaboration with Israel Aerospace Industries (IAI). The first squadrons were declared fully operational only in 2021.

The missile is designed to provide the armed forces with air defence capability against a variety of aerial threats at medium ranges. Each MRSAM weapon system comprises of one command and control system, one tracking radar, missiles, and mobile launcher systems. The mobile launcher is used to transport, emplace and launch up to eight canisterised missiles in two stacks. It can fire the missiles in single or ripple firing modes from the vertical firing position.

The combat management system simplifies the process of engaging a variety of threats. It identifies and tracks the threat using tracking radar. The system calculates the distance between the target and the launcher, and then determines if the identified target is a friend or a foe. The target information is, then, transmitted to the mobile launcher.

The weapon is 4.5 metres long, weighs approximately 276kg, and is equipped with canards and fins for control and manoeuvrability. The MRSAM missile is equipped with an advanced active radar radio frequency (RF) seeker, advanced rotating phased array radar and a bidirectional data link. The RF seeker, located in the front section of the missile, is used to detect moving targets in all weather conditions. The phased array radar provides a high-quality air situation picture, while the bidirectional data link is used for relaying midcourse guidance and target information to the missile. The missile's explosive warhead, featuring a self-destruct fuse, provides high probability of kill against enemy targets with minimal collateral damage.

The MRSAM surface-to-air missile is powered by a dual-pulse solid propulsion system developed by DRDO. The propulsion system, coupled with a thrust vector control system, allows the missile to move at a maximum speed of Mach 2. The weapon has the ability to engage multiple targets simultaneously at ranges of 70km.

The Akash-1 SAM

The Akash SAM, which equips eight squadrons, is the first indigenous SAM system to be inducted into the IAF. Each Akash battery consists of four self-propelled Launchers (3 Akash SAMs each), a Battery Level Radar – the Rajendra – and a command post (Battery Control Centre). Two batteries are deployed as a squadron (Air Force), while up to four form an Akash Group (army configuration).

Test-launch of an Akash NG surface-to-air missile. (DRDO)

A closer look at an Akash-launcher of the IAF. (DRDO)

In both configurations, an extra Group Control Centre (GCC) is added, which acts as the Command and Control HQ of the squadron or group. Based on a single mobile platform, GCC establishes links with Battery Control Centres and conducts air defence operations in coordination with air defence set up in a zone of operations. For early warning, the GCC relies on the Central Acquisition Radar (CAR). However, individual batteries can also be deployed with the cheaper, 2D BSR (Battery Surveillance Radar) with a range of over 100km.

Akash has an advanced automated functioning capability. The 3D CAR automatically starts tracking targets at a distance of around 150km providing early warning to the system and operators. The target track information is transferred to GCC. GCC automatically classifies the target. BSR starts tracking targets around a range of 100 km. This data is transferred to GCC. The GCC performs multi-radar tracking and carries out track correlation and data fusion. Target position information is sent to the BLR which uses this information to acquire the targets.

The BCC which can engage a target(s) from the selected list at the earliest point of time is assigned the target in real time by the GCC. The availability of missiles and the health of the missiles are also taken into consideration during this process. Fresh targets are assigned as and when intercepts with assigned targets are completed. A single shot kill probability of 88 percent has been achieved by the system taking into consideration various parameters of the sensors, guidance command, missile capabilities and kill zone computations.

Each Akash battery can engage up to four targets simultaneously. Each battery has four launchers with three missiles each, with each Rajendra able to guide eight missiles in total, with a maximum of two missiles per target. Up to a maximum of four targets can be engaged simultaneously by a typical battery with a single Rajendra if one (or two) missiles is allotted per target. A single Akash missile has an 88 percent Probability of Kill. Two missiles can be fired, five seconds apart, to raise the Probability of Kill to 98.5 percent. Communications between the various vehicles are a combination of wireless and wired links. The entire system is designed to be set up quickly and to be highly mobile for high survivability. The Akash system can be deployed by rail, road or air.

The Akash 1S and Akash NG

As noted earlier, seven squadrons of an upgraded version of the Akash SAM – the Akash-1S – have been approved and, given the positive results of trials of the Akash-1S, the induction of these seven squadrons will mean that the IAF will operate no fewer than 15 squadrons of the Akash with perhaps future inductions of other Akash variants to come. The first of these, the Akash NG, with an enhanced range of between 40km and 70km, began its testing process in 2021 and it is believed that this new version will become the definitive variant of the Akash and possibly replace the entire remaining force of Pechora SAMs. It would appear that the Akash NG is more compact than the original Akash system but with improvement to guidance and propellants to give it a much greater performance as compared to the original. More importantly, the system is much more mobile, thus enhancing its flexibility.

The QRSAM

The QRSAM is planned to supplement and later replace the Spyder and Osa-AKM SAM-systems. (DRDO)

Air Defence Tactical Control Radar of the IAF. (DRDO)

Continuing on the path of inducting more indigenous systems, the IAF can also look forward to a new Indian quick-reaction SAM (QRSAM) entering its inventory in the near future. Tests of the QRSAM began in 2017 and development is continuing. The Mach 1.8 QRSAM will have a kill-zone of between 3km and 30km in range, from 30 metres to 6km in altitude, and 360 degrees in azimuth.

For the IAF, a QRSAM squadron will comprise a Command Post Vehicle (RCPV), one S-band 90km-range air-defence tactical control radar (ADTCR) for airspace surveillance, two to three flights coordinated by a Battery Command Post Vehicle (BCPV), and a 120km-range C-band active phased-array Battery Surveillance Radar (BSR). Each troop in turn will comprise an X-band 80km-range active phased-array Battery Multi-Function Radar (BMFR), plus a 16km-range optronic fire-control system, and four Missile Launch Vehicles (MLV), each of which will carry six canister-encased missiles.

The S-400

Into this mix, five squadrons of S-400 SAMs are to be added. However, as will be detailed

The MRSAM is planned to equip a total of nine IAF squadrons. (PIB)

later in this chapter, the procurement of the S-400 may have more to do with India's quest for Ballistic Missile Defences as much as for the need for a long-range SAM.

For long-range SAMs, India has placed much emphasis on the acquisition of five squadrons of S-400 SAMs but there are persistent reports of a DRDO project for an extra-long-range SAM with a performance similar to that of the S-400. Another possibility is that the MRSAM receives an additional booster to give it a range of over 150km.

The long range of the S-400 in the anti-aircraft role allows a level of extra-territorial interception hitherto impossible with shorter-ranged systems. This would be particularly effective against hostile AEW or reconnaissance platforms or would serve to severely disrupt assembling strike packages well into enemy airspace.

The S-400 has four different types of missiles with ranges between 40km, 100km, 200km and 400km. It can also be deployed in a very short time and with its 92N6E electronically-steered phased array radar, it is highly resistant to electronic jamming. The S-400 also offers a limited ballistic missile intercept capability, augmenting Indian efforts in that direction.

The first of five squadrons will enter service at the end of 2021 and, despite the threat of US sanctions through the ill-conceived legislation designed to deter Russian arms imports, the Indian government and air force are intent on inducting the S-400 which will represent a quantum leap in the IAF's long-range SAM inventory.

Ground-based Air Defences – IAF AAA Gun Units

The IAF is for the first time contemplating raising AD gun units under a proposed Close-In-Weapon System (CIWS) project for 244 AD guns. A limited tender has already been floated for the CIWS programme with Bharat Forge Limited, Punj Lloyd, Tata Power SED, Larsen & Toubro, Reliance Defence, and Mahindra Defence Systems, as well as the state-controlled Bharat Electronics Limited and Ordnance Factory Board being participants.

The IAF has identified the combination of these CIWS QRSAMs, Akash SAMs, MRSAMs and LRSAMs of the S-400 type as being required to complete its air defence requirements. As against these requirements some progress has been made for the procurement of all four types. However, the numbers of systems of each type being procured are somewhat limited as of now. Future orders for the Akash, MRSAM and S-400 should be expected as a matter of urgency.

Indigenisation Rules the Roost

It is of interest that the IAF has clearly placed its faith in indigenous systems. Even the MRSAM, soon to be IAF's pride, is a joint Indian-Israeli system. This, alongside the Akash and QRSAM systems plus the indigenous upgrades to legacy AD guns, clearly shows that the IAF is fully supportive of Indian capabilities and Indian industry. This extends to radars where indigenous developments are set to completely replace all legacy radars in IAF service. This will, inevitably, lead to a dramatic improvement in India's Air Defences.

Indian BMD Network and the Role of the IACCCS

Dealing with manned aircraft and cruise missiles is well within the technological capacity of India's existing systems which have undergone periodic upgrades as needed. However, the need for a BMD system – encompassing both exo-atmospheric and endo-atmospheric interception assets – presents a completely different challenge and has presented the IAF and DRDO with a number of technological and practical constraints. At the outset, it has to be reiterated that to date, India has not deployed any BMD systems and its capability is still nascent though steady progress has been made. Few efforts have been made to analyse the progress made to date and link it to other air defence developments. One of the few to have undertaken this is Prasun Sengupta.

Prasun Sengupta suggests that the most challenging and contentious part of the IACCCS' implementation roadmap is the two-tier BMD component. He notes that while the ground-based, airborne and space-based tools required for giving early warning of inbound hostile ballistic/cruise missiles are already being acquired from both indigenous sources and abroad (primarily Israel), the process to acquire and deploy an interception capability – the active 'hard-kill' component of anti-ballistic missiles and their fire-control systems – is going to take some considerable time. At the current rate of progress, it is unlikely that the initial components of such a two-tier BMD network, comprising both endo-atmospheric and exo-atmospheric missile interceptors, are unlikely to be commissioned before 2022. It is probable that the initial Indian BMD system will be comprised of a mixture of imported and domestic BMD interceptors and radars.

At present, for fire-control purposes the BMD system uses ELTA Systems-built EL/M-2080 'Green Pine' ground-based active phased-array L-band long-range tracking radar (LRTR). Two of these radars were supplied in late 2001 under the US$50 million 'Project Sword Fish' to the DRDO by the ELTA Systems Group subsidiary of Israel Aerospace Industries.[14] This has led to some confusion as some have assumed the name 'Sword Fish' to be a DRDO-developed radar. India did write three million lines of software code for the Battle

Management/Command, Control, Communications & Intelligence (BM/C³I) centre, the hub of software and hardware systems.

Furthermore, transmission links to the interceptor missile are based on jam-proof CDMA technology with multiple data transmission links which have been set up so that if one is jammed the others could function. In addition, Sengupta notes that Israeli inputs were sought and obtained for designing and fabricating the BM/C³I centre, which not only acts as the DRDO's primary BMD engagement simulator, but is also being used for evolving BM/C³I concepts, for defining BMD goals and developing BMD doctrine, for evaluating candidate systems architectures, for serving as the principal prototyping-cum-validation tool for the BMD's BM/C³I algorithms, and for defining the human role in the BMD battle. The BMD's endo-atmospheric element makes use of the THALES Raytheon-supplied S-band Master-A engagement radar.

Ballistic Missile Defences for India: Interceptor Missiles

India's efforts to obtain a ballistic missile interception capability date back to between 1996 and 1998. News began leaking out about the deployment from 1998 onwards of an Anti-tactical Ballistic Missile screen. This system was to comprise the Russian S-300V ATBM (SA-12) and India's own 'Akash' missile which has a considerable ATBM capability. In March 1997, the Indian press confirmed these reports, stating that one S-300V squadron was being purchased, with more to come in the future. These reports proved to be incorrect and as far as is known no variant of the S-300 has entered Indian service. The S-300 saga illustrates some of the many pitfalls of Indian defence reporting and coverage where apparently definitive reports of acquisitions turn out to be completely false. This has the dual effect of confusing analysis while simultaneously decreasing the credibility of analysts who, in good faith, believe such media reports.

India's quest for a BMD system has had many false starts and while still very much a work in progress has made some tangible steps towards the development of a viable system. Early efforts focused on evaluating the potential of the Akash SAM to intercept short-range ballistic missiles. To date this remains a potential development but has never been demonstrated. Other efforts included stillborn attempts to acquire S-300 PMU-1 and S-300V system from Russia, as noted above, and Arrow-2 systems from Israel.

More recently, there has been a more determined effort to acquire five squadrons of S-400 SAMs from Russia which do have a potential BMD role, especially in the variants sought by India. The contract was signed in 2018 and the first systems were delivered in 2022.[15]

India's indigenous efforts at developing a BMD system has had the misfortune of being plagued by a series of misstatements by DRDO officials which claimed deployment readiness when it was evident that the Indian system was more of a proof-of-concept technology demonstration until later versions of its endo- and exo-atmospheric interceptors became available.[16]

DRDO itself has acknowledged that its first efforts were not the definitive versions of the interceptor missiles and as such it was surprising to see claims of the system being deployable when this was patently not so. Such unfortunate statements have had the effect of obfuscating DRDO's genuine progress.

C4ISR

DRDO's overall BMD C4ISR architecture is intended to consist of both over the horizon and X-band fire control radars which detect and track incoming missiles, a mission control centre (MCC) that fuses input (which may also come from satellite-based sensors), processes it and then sends orders for engagement to launch

Exo-atmospheric interceptor of the PDV. (DRDO)

control centres (LCCs) situated up to a 1,000 km away via mobile communication terminals (MCTs). According to Saurav Jha, the LCCs then orchestrates the final launch sequence with the mobile interceptor sitting nearby.

Repeated tests of the two-tier system, including the latest PDV test, has given enough confidence to DRDO to recommend the freezing of the current configuration for Phase 1. Both the radars and the LCCs receive and send information via target update transmitters (TUTs) based on CDMA technology. According to Jha, the MCTs of the MCC are themselves connected via an IP wide area network; data-links for the entire setup also include fibre-optic communication channels and line-of-sight relays.[17]

A deployed Indian BMD system would consist of several launch vehicles, radars, LCCs and the MCC. Using a secure communication system, the MCC and perhaps secondary sites would be used to link a geographically widespread network. The MCC is the heart of the BMD system. It is a software intensive component which receives information from radars and satellites which is then processed by ten simultaneously running computers. The MCC is then connected to all other elements of the system through a WAN. The MCC would perform tasks such as target classification, target assignment and interception success assessment deciding the number of interceptors required for the target for an assured kill. After performing all these functions, the MCC assigns the target to the LCC of a launch battery. The LCC then computes the necessary time required and optimal window for launching the interceptor based upon information received from a radar based on the speed, altitude and flight path of the target.[18]

While experimental versions of tracking and fire-control radars are available for testing, India would need to establish production facilities for these radars. To date, there is little indication that any progress has been made in this regard, reflecting either a desire for a refined system to be finalised for production or concern that these may still need to be imported. Currently, India uses the imported radars mentioned earlier and has not placed future orders for additional systems.

Endo-Atmospheric Interceptors – the AAD

The Advanced Air Defence (AAD) system is an anti-ballistic missile designed to intercept incoming ballistic missiles in the endo-

A test launch of an AAD from a mobile transporter-erector-launcher (TEL). (DRDO)

atmosphere at altitudes of between 15km and 30km. It is a single-stage, solid-fuelled missile. Its guidance is based on an inertial navigation system, with midcourse updates from ground-based radar and it has an active radar homing in the terminal phase. It is 7.5m (25ft) tall, weighs around 1.2t (1.2 long tons; 1.3 short tons) and a diameter of less than 0.5m (1ft 8in).[19]

Compared to the exo-atmospheric systems, the AAD has been tested often and against relatively realistic targets. On 6 December 2007, AAD successfully intercepted a modified Prithvi-II missile acting as an incoming enemy ballistic missile target. The endo-atmospheric interception was carried out at an altitude of 15km. The AAD was launched when the Prithvi reached an apogee of 110km (68 miles). The AAD, with the help of midcourse updates and its terminal seeker, manoeuvred itself towards the target and scored a hit on the target at an altitude of 15km and at a speed of Mach 4. Ground and ship-based radars detected the formation of a large number of tracks, signifying that the target had broken into multiple pieces and these were confirmed by thermal cameras located on Wheeler Island. This was followed on 26 July 2010, when the AAD was successfully test-fired from the Integrated Test Range (ITR) at Wheeler Island.

On 6 March 2011, India launched its indigenously-developed interceptor missile from the Odisha coast. India successfully test-fired its interceptor missile which destroyed a 'hostile' target ballistic missile, a modified Prithvi, at an altitude of 16km over the Bay of Bengal. The interceptor, Advanced Air Defence (AAD) missile positioned at Wheeler Island, about 70km across the sea from Chandipur, received signals from tracking radars installed along the coastline and travelled through the sky at a speed of 4.5 Mach to destroy it. The interceptor missile had its own mobile launcher, secure data link for interception, independent tracking and homing capabilities and sophisticated radars.[20] Another test was conducted on 10 February 2012.

Finally, on 23 November 2012, India again successfully test-fired its home-made supersonic Advanced Air Defence (AAD) interceptor missile from a defence base off the coast of the eastern state of Odisha. The test-firing was part of India's efforts to create a missile defence shield against incoming enemy missiles. The AAD interceptor missile, which was fired from Wheeler Island off the Odisha coast, successfully destroyed mid-air an incoming ballistic missile launched from the Integrated Test Range in Chandipur, about 70 km from Wheeler Island.[21]

The Upgraded AAD: The Definitive Endo-atmospheric Interceptor

On 6 April 2015 an improved AAD was tested. The missile was launched from a canister for the first time and the composite rocket motor fired successfully. The upgraded AAD improves over the AAD as it has a bigger warhead, improved manoeuvrability and a higher hit-probability. The test, however, was unsuccessful. As the missile was in flight, one of the sub systems malfunctioned causing the interceptor to veer away from the flight path resulting in the failure of the mission.[22]

On 22 November 2015 the upgraded AAD was successfully tested. The anti-ballistic missile took off at 9:40 a.m. from the A.P.J. Abdul Kalam (Wheeler) Island soon after it received the command to waylay and destroy an incoming electronically-simulated target missile. Conditions similar to the launch of a target missile from Balasore were simulated electronically and upon receiving its coordinates, the interceptor missile, travelling at supersonic speed, engaged and destroyed the virtual target in mid-flight.[23] The first test against a live target took place on 15 May 2016. During this test, DRDO officially reported that the UPGRADED AAD interceptor intercepted and destroyed a Prithvi ballistic missile fired from a ship.[24] However, some reports suggest that the intercept failed with the UPGRADED AAD failing to launch, much less intercept the target.[25]

A further test took place on 1 March 2017 when the upgraded AAD interceptor achieved a terminal intercept of the target missile at an altitude of 15 km. A further test on 28 December 2017 achieved a hit-to-kill intercept at an altitude of 15km.[26]

The upgraded AAD configuration for these tests were similar to what would be offered for production and deployment. This test achieved both an actual intercept of an incoming ballistic missile target, as well as an explosive intercept, as the pre-fragmented warhead on-board the upgraded AAD was detonated using the missile's radio proximity fuse (RPF).[27] The upgraded AAD was initially guided by its on-board inertial navigation system (INS) which received continuous updates about the incoming target's trajectory from ground-based radars through a secure data link. Subsequently, a radio frequency seeker in the upgraded AAD's nose cone section tracked the target while an intercept course was plotted by its on-board computer.[28]

To summarise, as compared to the exo-atmospheric interceptors, development of the AAD and the upgraded AAD endo-atmospheric interceptor missiles has witnessed greater urgency, with the AAD being test-fired on 6 December 2007, 6 March 2009, 15 March 2010, 26 July 2010, 6 March 2011, 10 February 2012 and 23 November 2012. As has been detailed above, upgraded AAD missile's test-firings commenced on 6 April 2015 and were followed by test firings on 23 November 2015, 15 May 2016, 1 March 2017 and 28 December 2017.

The version of the Long Range Tracking Radar used in the May 2017 UPGRADED AAD mission is an L-band array that can track a ballistic target with a radar cross section (RCS) of 0.1 square metres from over 1,500km away. The MFCR, which is a S-band array, has a tracking range of over 370km for a target with a RCS of 0.3 square metres. These detection figures may be sufficient to detect a re-entry vehicle but this is still a matter for some conjecture and as such any estimation of its effectiveness against RVs is still somewhat theoretical, though probable.

Exo-Atmospheric Interceptors

India's efforts to develop an exo-atmospheric interceptor have been markedly slower than those made in respect of their endo-atmospheric interceptors. This has made any claim of either phase 1 or phase 2 of India's BMD system being ready for deployment somewhat incredible. To date, only four tests of exo-atmospheric interceptors have been carried out, two of which were with what can best be described as a proof of concept demonstrator, based on the liquid-fuelled Prithvi SSM. This is an impractical system for deployment as an interceptor missile and as such India has opted to move away from that towards a potentially more efficient and effective system. A critical aspect for successful exo-atmospheric interception is target detection. Sufficient warning time must be available for a successful intercept. India is yet to develop or deploy a missile early warning radar suitable for the detection of longer-ranged missiles and lacks space-based radars. Until such time as these systems are developed and deployed, India's exo-atmospheric interception capability will remain somewhat limited.

India's requirements for an exo-atmospheric interceptor have acquired a degree of urgency with the Indian Air Force identifying multiple targets which need to be protected. To this end, India and Russia agreed the sale of five S-400 SAM/TMD systems to India specifically for the purpose of providing a degree of protection for vital areas and potential targets within India. The procurement of the S-400 would provide India with a considerable degree of exo-atmospheric interception capability as one version of the system is capable of intercepts at altitudes of up to 185km, approximating the performance of the US THAAD system.

India's efforts have not been insubstantial in this regard with intercepts achieved at altitudes of 47km, 75km and most recently 97km, with the existing PDV missile purportedly having the capability to intercept targets at altitudes exceeding 150km and itself being the precursor for a second phase of development aimed at producing an interceptor with an intercept ceiling of 300km.

Yet, India's progress in this respect has been slow. Only four tests have been conducted of exo-atmospheric interceptor missiles over an 11-year period. This could be due to the greater technological challenges faced in the development of these missiles or it could also be due to the greater priority assigned to the AAD and UPGRADED AAD endo-atmospheric interceptor missiles.

The Prithvi Air Defence Missile

The first test of India's indigenous BMD was the test of the PAD (Prithvi Air Defence) missile in November 2006. During this exercise, a PAD missile successfully intercepted a modified Prithvi-II missile at an altitude of 47km. The Prithvi-II ballistic missile was modified successfully to mimic the trajectory of M-11 missiles.

A second test took place in March 2009, when India conducted a test using ship-launched Dhanush missile (naval version of the Prithvi missile) as the target simulating a missile with a range of 1,500 km. Swordfish (the LRTR) radar was used for tracking and destroyed using a PAD missile at an altitude of 75km.[29]

Following the 2009 test, DRDO made claims that India had completed phase 1 of the project capable of engaging targets at a 2,000 km range and that it was in the process of developing missiles to tackle targets at a 5,000 km range. These claims ought not to be taken seriously considering that India's exo-atmospheric interception systems are as yet seriously deficient. While some capability has been demonstrated, it is impractical for India to consider the deployment of an *ad hoc* system that would be costly and relatively ineffective.

The exo-atmospheric stage of the Indian BMD system went through two phases – the first where the exo-atmospheric interceptor was effectively a re-engineered Prithvi SSM and the second where a new and much more capable exo-atmospheric interceptor was fielded. Initially, as stated above, the two-tiered BMD system consisted of the PAD, which demonstrated the ability to intercept missiles at altitudes of 47–75km. To date, the Indian system has been tested on target missiles based on the Prithvi SSM modified for longer ranges and smaller radar cross-sections. For exo-atmospheric interceptions, it is essential that this target missile be replaced by one with a separating re-entry vehicle to properly test the interceptor missile.

The PAD, using a liquid-fuelled interceptor was impractical from a fuelling and storage perspective. The corrosive liquid fuel used in the Prithvi missile makes storage of a fully-fuelled missile impossible beyond a five year period. Storing the missiles in an unfuelled state is not merely impractical but defeating for the purpose of a BMD defence system which needs to be ready to intercept a target at short notice and which may need to be stored in a ready-to-use form for prolonged periods. The liquid-fuelled PAD cannot, for example, be considered for canister launch. To this end, India has moved forward with a much more advanced and practical system.

The PDV

For exo-atmospheric interceptions, the PAD has been replaced by the much more capable PDV missile which is intended to carry out interceptions up to an altitude of 150km. On 27 April 2014, the first PDV was successfully tested against an electronic target. This simulation seemed to be intended to validate the parameters of the interceptor missile rather than to achieve an actual interception and as such has to be viewed as a test launch rather than an actual engagement.[30] Furthermore, it was used to validate the interceptor's integration with the detection, tracking and automated launch control systems.

This was followed by an actual intercept on 11 February 2017, when a target missile was engaged in a successful hit-to-kill intercept at an altitude of 97km. While this is a considerable achievement, unlike earlier instances when claims of near deployment were being made, DRDO issued a much more nuanced statement indicating the requirement for more tests and a more effective detection system were emphasised, though the capability of the PDV was undeniable.[31]

Saurav Jha, in a detailed report on the interception, noted that the hit-to-kill (HTK) interception achieved by the exo-atmospheric PDV was a major achievement for the Indian BMD programme. By achieving an exo-atmospheric intercept at an altitude of 97 km, the 11 February test validated, among other things, an improved guidance algorithm used for this test-mission, as the incoming missile target had deviated significantly from what would have allowed an intercept along a typical minimum energy trajectory (MET). He noted that the PDV interceptor had to hit the target at the far end of its engagement envelope and at a lower altitude than what a 'standard' MET intercept would have entailed.[32]

This test, Jha further notes, could indicate the maturity of the on-board imaging infrared seeker (IIR), the responsiveness of the divert and attitude control system (DACS) used by the PDV's kinetic kill vehicle (KKV), as well as the sensor fusion achieved by various tracking systems involved in the mission. The February 2017 test, unlike the 2014 one, was used to prove the efficacy of the KKV used by PDV by destroying an actual incoming warhead in a HTK mission.

Compared to the PAD, which Jha describes essentially a high endo-atmospheric system with a maximum ceiling of around 85 km, the PDV is a genuine exo-atmospheric interceptor capable of destroying targets at altitudes of up to 150 km.[33] In addition, instead of using a radio frequency (RF) seeker like the PAD, the PDV uses a strap-down IIR seeker developed by DRDO's Research Centre Imarat (RCI) with a 128 x 128 focal plane array.[34] This IIR seeker enables the interceptor to discriminate against missile warheads and decoys. The PDV's inertial guidance package includes a ring-laser gyroscope (RLG) which enables its solid-fuelled booster to move towards the estimated point of interception as calculated by ground-based radars, until the KKV is released and the integral IIR seeker takes over in the terminal phase to track the RV. After which, the KKV steers itself continuously to plot a collision course with the incoming RV. In the February 2017 test, Jha reports that the KKV managed to smash right into the central portion of the RV.[35]

Acknowledging that a new interceptor in the class PDV requires a new MRBM class target missile for effective trials, DRDO developed a new target missile. This used a two-stage target, comprising a new solid-fuelled second stage that sits atop a liquid-fuelled Prithvi booster first stage. This target missile successfully simulates the 3–4km/sec re-entry speeds of a 'hostile' ballistic missile approaching from more than 2,000km away. This new target missile provides a considerably more realistic challenge for the PDV as compared to that used in the PDV and AAD tests which would invariably have been limited in range and performance.

Prasun Sengupta notes that the PDV will take at least a decade to mature. It is designed to intercept MRBMs (with atmospheric re-entry speeds of 5km/second) more than 500km away at an altitude of 150km. The PDV will cruise at Mach 5 but will be required to attain a peak terminal speed of Mach 11 – made possible by the divert thruster placed on top of the second stage. This divert thruster will generate high lateral acceleration for the 'end-game'. Both the warhead and divert thruster will be fired simultaneously towards the target once they are within the acquisition range of the PDV's imaging infrared seeker (IIR).[36]

A critical aspect of a layered approach to BMD is the verification of the modularity of the systems involved. This would allow the use of different interceptors – with varying interception parameters – with the same command and control network (C2) which would inevitably lead to cost-effectiveness through interoperability. The February 2017 PDV test validated the successful integration of the PDV interceptor with DRDO's ground based automated response network that has been used for the AAD and UPGRADED AAD interceptors and which forms the mainstay of the two-tier BMD concept.[37]

The PDV, as an exo-atmospheric interceptor, is larger and has more on-board fuel than the AAD and its successor, the UPGRADED AAD endo-atmospheric interceptor. Unlike the PAD, the PDV booster uses solid propellants which have high burn-rates and can function effectively in temperatures ranging from minus 40 degrees to 50 degrees Celsius. This has required special casting for the propellants developed by DRDO's High Energy Material Research laboratory (HEMRL) which has required significant improvements in metallurgy.[38]

Recognizing the inherent deficiencies in the PAD, which cannot be stored in a ready to launch role, the PDV is also designed to have a shelf life of 10 to 15 years before overhaul and its propulsion systems have a high margin for safety while retaining quick reaction capability. This makes the system suited, when fully developed, to being deployed in a ready to launch mode, slaved to an automated launch mechanism. Furthermore, PDV's propulsion systems are robust enough to withstand being moved, facilitating deployment of the system as a mobile platform. Jha notes that the DACS onboard PDV's KKV is fuelled by hypergolic propellants, with high thruster valves which can precisely control the flow of propellant to the rocket engines used for KKV steering.[39]

While its development cycle is still a work in progress, once the development of PDV is successfully complete, it would signal the maturing of Phase 1 of India's BMD programme which is designed to provide credible capability against theatre ballistic missiles (TBM) launched from up to 2,000km away. Moreover, as Jha has pointed out, the PDV gives an indication that India has low earth orbit (LEO) capabilities through the PDV. As the PDV is but a step towards another and more capable exo-atmospheric interceptor with the capability to intercept re-entry vehicles at altitudes exceeding 300km, India has developed the building blocks to deploy an anti-satellite system. This was tested in 2019.[40]

Deployment of a BMD System is Getting Closer

In pursuance of a BMD network, the first steps have been taken to secure clearance for the deployment of a limited BMD system around Delhi with the IAF and DRDO approaching the government for such a system which will be done as part of a broader integrated, national air defence network which will fully integrate army and navy radars and SAMs with the IAF's network.[41] Initial deployment of the AAD and PDV systems would be very limited but once sanction is given, it would mark an important first step. Sanction for this has not yet been received but the need for a BMD system has been long been deemed necessary and as such, it may be expected fairly soon.

6
THE INDIAN AIR FORCE AND NUCLEAR WEAPONS

With a force of some 260 Su-30MKIs, 36 Rafales, 49 Mirage 2000s and over 110 Jaguars, the Indian Air Force has many options for manned aircraft delivering nuclear gravity bombs. The weight, yield and dimensions of Indian gravity and/or glide bombs are not known, however it might be anticipated that Indian weapons range between 250kg and 1,000kg and can have a yield of between 15 kilotons and maybe as much as 200 kilotons – the latter being fusion-boosted-fission weapons.

India's nuclear arsenal is shrouded in excessive secrecy. While some of this is inevitable, the complete lack of information on the Indian nuclear weapons inventory, delivery options and command and control is somewhat disconcerting. However, as of 2017, it can be safely said that the Indian arsenal – under the control of the Strategic Forces Command (SFC) is at present land-based and consists of air-delivered weapons and land-based ballistic missiles. While India has made significant strides in developing a submarine-based deterrent, the INS Arihant and her K-4 missiles are a considerable distance away from attaining operational status. In the medium term, these systems will become relevant but at this juncture, India would plan its nuclear deployment and delivery systems around manned aircraft and a mix of short, medium and long-range ballistic missiles with cruise missiles – air, land and sea launched – being very probable.

India's Nuclear Weapons – From Testing to Weaponisation

India's first nuclear weapons test was on 18 May 1974. The device tested at Pokhran had a yield of some 12 kilotons and used the implosion principle. It is not known exactly how much plutonium was used for the test, but some informed Indian guesstimates indicated that it was about 10kg – about one year's output from the 'Cirus' reactor. This figure includes some plutonium lost during the machining of the core of the bomb. This leaves about 6–8 kg in the device itself. The yield of the device was questioned with estimates ranging from the official 12 kilotons to a low of two kilotons, with the latter estimate being dismissed. However, P.K. Iyengar believed that the yield was between 8 and10 kilotons based on radiochemical analysis from the shaft.[1]

The Pokhran device was exploded at a depth of 107 metres after being placed in an L-shaped hole. It was long believed that the device tested was a 'crude' bomb. However, in a 1994 magazine interview, the head of India's Atomic Energy Commission, Dr. Rajagopala Chidambaram, stated that India was confident of the design prior to the test and almost boasted about how 'good' the Indian 'bomb' was.[2]

It is interesting to note that Dr. Chidambaram, who was closely associated with designing the Pokharan device, used the word 'bomb' instead of the more widely used 'device'. Moreover, some observers point out that the 1974 device was small enough to fit down a metal pipe and hence would be small enough to fit into a bomb casing. It may therefore be argued that the 1974 test was more of weapon than a mere device.

Indecision and Tentative Steps

Subsequently, India's first effort to formulate a nuclear policy and to ascertain how to effectively implement it began in November 1985 and was aimed at answering questions posed by Rajiv Gandhi. Comprising representatives from the three services (Navy Chief of Staff Admiral Tahliani, Army Vice Chief of Staff General K. Sundarji, Deputy Chief of Air Staff John Greene) plus influential representatives from BARC, DRDO and the AEC, joined India's most prominent strategic analyst at the time, K. Subrahmanyam. The outcome of the group's deliberations was to recommend building a minimum deterrent force of 70–100 warheads with a strict no-first-use policy.[3] It appears even at this time, a hastily cobbled together system for weapons delivery was available.

While no direct action was taken in respect of the recommendations from that group, in 1986 Gandhi instructed V.S. Arunachalam of DRDO to develop a delivery system with suitable control and security measures and improved reliability to replace the stopgap system developed two years earlier. Arunachalam recruited K. Santhanam as technology adviser on the project and the bomb system was codenamed 'New Armament Breaking Ammunition and Projectile', or NABAP.[4] Arunachalam went further and studied the basis for a basic command and control network that would be survivable. Based on his work, Arun Singh was ordered to set up a national command post at a secure location near to New Delhi and developed a semi-operational air-delivered deterrent.[5]

Unlike previous efforts, the IAF was involved and this perhaps revealed flaws in the weapon's development with the weapon designed being decidedly unsuitable for delivery from the Jaguar aircraft selected for weapons delivery, only two inches of ground clearance being available.[6] By late 1986 the Air Force rejected the Jaguar as unsuitable, and efforts switched to integrating the bomb with the recently acquired Mirage 2000 and this led to the eventual adoption of this aircraft as its primary weapons delivery system until missiles became available.

During the tensions over Kashmir between 1989–90, India became acutely aware of its deficiencies in respect of nuclear command and control. To this end, the Arun Singh Committee was established to plan and prepared to plan India's nuclear emergency response measures in the event of nuclear escalation. This committee prepared a series of emergency response procedures and established command and control mechanisms but did not deal with operational planning.

If Krishnaswamy Subramanyam, who was a participant, is to be believed, the committee's 'only specific recommendation', was to 'to create separate storage for missiles and warheads…what should be the drill for them being brought together…and then…the communications from command and control.'[7]

George Perkovich claims that, 'The group called for designating air force units to receive nuclear devices and deliver them according to previously prepared orders that base commanders would possess under seal.'[8]

The 1990 crisis provided an additional spark for the redesigning of India's air-delivered bomb to ensure comfortable carriage under a Mirage 2000. This ultimately led to the certification of the air-delivery platform and weapon from a Mirage 2000 in 1994–95.[9] After this step, in 1995, then Prime Minister Narasimha Rao approved the implementation of dispersal and concealment routines and procedures that had planned for safeguarding both the fissile cores and non-fissile trigger assemblies from a pre-emptive attack.[10]

The First Weapons

Indications are that the first Indian nuclear weapons design had a mass of about 1,000kg with a yield of 12–15 kilotons. However, subsequently, perhaps by 1982 (when rumours of a fresh round of nuclear tests were being circulated), the said weapon had been scaled down to a more manageable mass of between 170kg and 200kg.[11] It appears that a 100 kiloton fission weapon was later produced for aerial delivery with a mass of 200 to 300kg.[12] If this is accurate, it would mean that India had perfected a relatively high-yield fission weapon with a relatively low mass for its class. It should also be noted that by the 1990s, fusion-boosted-fission weapons had become available for air delivery.

While India had developed a stockpile of some 15 to 20 nuclear gravity bombs in the period 1974–1985 – all of which were in the 15–20 kiloton range – India became an effective nuclear weapons state in May 1994. It might have been thought that this milestone would have been achieved earlier, and indeed it was certainly possible for this to be done. However, mating of the nuclear weapons with the delivery aircraft proved to be challenging.

India's first deployed air-delivered nuclear weapons were earmarked for use from the Jaguar strike aircraft. One would have assumed that in the 1980s and into the early 1990s, the Jaguar with an excellent navigation-attack system and a very good payload capacity would have been an ideal choice. However, the type's less than stellar flight performance combined with practical problems of ground clearance led to the Mirage 2000 being chosen to perform this task, with active trials of the system being undertaken in 1994.

Development of India's air-launched deterrent was somewhat chaotic. Though at the urging of General Sundarji, Prime Minister Rajiv Gandhi had instructed DRDO to start development of rugged, miniaturised, safer nuclear weapons and to develop more reliable components and subsystems for what was termed as and intention to 'keep the country's nuclear capability at least at a minimum state of readiness.'[13]

While some argue that it stopped short of ordering the building of a weapon or integrating it into a delivery platform, this could be debated ad nauseum as that period of Indian nuclear history is still somewhat unclear. However, eventually, in 1989 weaponisation was approved after Rajiv Gandhi's disarmament plan for the globe met with abject failure.[14] In the wake of the failure of his global disarmament plan there were menacing Indian intelligence reports, which concluded in March 1988 that 'Pakistan was in possession of at least three nuclear devices of 15–20 kiloton yield.'[15]

No missiles were yet in the offing and the instructions from India's Prime Minister were to concentrate on the development and certification of air deliverable weapons and the creation of an aircraft delivery package that was safe and reliable.[16] This was done but it took several years and a new Prime Minister – in the form of P.V. Narasimha Rao – to complete this task.

Initial tests with the Jaguar IS – India's principal deep-penetration strike aircraft – found that the ARDE (Armament and Research Development Establishment) designed bomb-pods were too heavy for the Jaguar and once fitted to the centreline pylon of the aircraft, the ground clearance of a mere two inches was deemed unsafe.[17] India then chose to use its Mirage 2000H/TH fleet for the nuclear delivery role.

It was with this aircraft that India finally completed its development of a fully combat ready system and deployable system for delivering nuclear weapons with the IAF conducting acceptance trials. The ARDE developed bomb case, and the TBRL (Terminal Ballistic Research Laboratory) developed implosion system were mated with a modified Mirage 2000 and successfully test dropped at Balasore using a toss-bombing technique. The bomb, minus its plutonium core, was fused for an airburst and released over the ocean. With this test, India has a reasonably reliable delivery system and by 1994, India now had an arsenal of at least a couple of dozen operational nuclear bombs.[18]

Whether true or not, it is widely believed that up to 1998, the air force was the only military service with any knowledge of the India's nuclear weapons programme and that was only because of its role in delivering the weapons.[19] While there was much unofficial knowledge of the fact that weapons were available for use, Gaurav Kampani noted that:

A senior Indian defense official privy to this effort disclosed that, until 1999, the air force had no idea (1) what types of weapons were available; (2) in how many numbers; and (3) what it was expected to do with the weapons. All the air force had was delivery capability in the form of a few modified Mirage 2000s.[20]

The 1998 Tests – Overt Weaponisation

India, after prolonged development, has now deployed a large and sophisticated nuclear arsenal. In 1999, however, assessments of its nuclear capabilities were still largely confined to views that only fission weapons in the form of free-fall bombs in the 15–20 kiloton range were available for deployment. With the benefit of hindsight, it is now quite well established that both India and Pakistan had deployed nuclear weapons in either the late 1980s or at the latest, the mid-1990s. What changed, however, was that in May 1998, India, then Pakistan, tested nuclear weapons, thus refining and experimenting with their designs as well as confirming their nuclear status.

In the period leading up to the 1998 tests, the Indian plutonium stockpile had been estimated to be sufficient for 85 weapons, though this might have been an exaggeration.[21] India had developed a stockpile of some 15 to 20 nuclear gravity bombs in the period 1974–1985 – all of which were in the 15–20 kiloton range – and became an effective nuclear weapons state in May 1994 when it tested its air delivery system using Mirage 2000 aircraft. Though this could have been achieved much earlier, and indeed it was certainly possible for this to be done, hesitancy was the hallmark of early Indian weaponisation. India had kept its weapons disassembled and in an un-mated state to its delivery systems, anticipating that 72 hours would be sufficient to fully operationalise its deterrent.

Known as 'Operation Shakti-98', the five tests had a recorded seismic magnitude of 5.0 +/- 0.4 on the Richter scale.[22] By 14 May, Indian seismologists at the Bhabha Atomic Research Centre (BARC) stated that an analysis of the data received at the Gauribidnaur Seismic Array indicated that the yield of the May 11 1998 blasts was around 55 kilotons.[23] They also indicated that the seismic waveform was very complex because the explosions were carried out simultaneously.

The first group of tests – Shakti I, II and III – consisted of a two-stage thermonuclear device with a yield of 43–45 kilotons, a lightweight fission device with a yield of 12–15 kilotons and a low-yield subkiloton device of 0.2 kilotons. These were placed in two shafts, one kilometre apart, and simultaneously detonated from a control located 3.5 km away.[24] Shakti IV and V had yields of 0.5 and 0.3 kilotons and were conducted to provide additional data for improved computer simulation of designs.

Weaponisation – the Kargil Watershed

Neither India nor Pakistan has publicly discussed its nuclear weapons deployment during the Kargil conflict of 1999. However, it is known that India readied four Prithvi SRBMs and one Agni IRBM for possible use in a retaliatory role in the event of the use of a Pakistani weapon, as well as preparing some of its Air Force Mirage 2000s to delivery free-fall bombs, bringing its arsenal to Readiness State 3 with warheads ready to be mated at short notice.[25]

In fact, Kargil was to be an eye-opening experience where the Indian nuclear deterrent took nearly a week to achieve any degree of capability – transitioning from a recessed deterrent to an employment mode – as opposed to an earlier calculation of 72 hours.[26] Indeed, Kampani notes that the air force was not ready until the end of June 1999 to commence nuclear operations against Pakistan if ordered.[27]

However, the IAF was not the only agency involved as DRDO had readied one Agni and four Prithvi missiles for the nuclear strike role against Pakistan with a trajectory for the former being worked out so that its discarded stages did not fall in Indian territory. Activation of this Readiness State 3, while taking longer than anticipated, marked the deployment of an operational air- and missile-based deterrent for India.[28]

That these preparations were in fact warranted, can be gleaned from a 2018 statement by Bruce Riedel, a former CIA analyst and a White House aide during the Kargil War, who indicated that the CIA's daily brief for 4 July 1999 indicated that Pakistan was preparing its nuclear weapons for deployment and possible use.[29] In the event, neither India nor Pakistan overtly threatened the use of their nuclear weapons during this conflict and it is possible that the plausible deniability sought by Pakistan plus India's response in not crossing the Line of Control were both important factors in moderating nuclear rhetoric.

Air-Delivered Weapons Today

At least one of India's three IAF Mirage 2000 squadrons was tasked with the nuclear strike role – although it does not seem that this was its exclusive role given the limited number of aircraft available. One might expect that the 36 Dassault Rafales may find themselves supplementing and thereafter supplanting the Mirages in the nuclear strike role.[30] It has also been speculated that the Su-30MKI fleet may have a nuclear strike role. This would not be unrealistic and the greater payload capacity offered by the Su-30MKI could eventually permit the carriage of air-launched nuclear-armed cruise missiles in the future. In fact, it has been suggested that the Su-30MKI has now supplanted the Mirage 2000 in the nuclear strike role.

It is of course preposterous to expect all Indian aircraft to be earmarked for the nuclear strike role. Rather, it is believed that certain squadrons have been given such a tasking in wartime with at least one squadron of each of the Su-30MKIs and Mirage 2000s being so employed. It might be expected that a degree of additional EMP protection would be provided to these aircraft. However, given the fact that it is highly unlikely that whole squadrons can be permanently assigned to the nuclear strike role without compromising India's ability to conduct air operations in the event of hostilities, determining which aircraft need to be given additional EMP shielding presents a challenge. It might be necessary for India to consider additional EMP shielding for a larger proportion of its air assets than would be necessitated by the number of nuclear weapons earmarked for aerial delivery.

India lacks a heavy bomber aircraft type equivalent to either the American B-52 and B-1B or the Russian Tu-22M3 and Tu-160 aircraft. It is interesting that India was offered the earlier Tu-22 as a replacement for the venerable Canberra bomber but for some reason declined the type – perhaps wisely as the aircraft proved less than successful in service. What is more surprising is that India has not moved towards obtaining an effective strike aircraft, perhaps such as the Su-34, to augment its multi-role aircraft which currently carry the dual responsibilities of conventional air combat tasks as

A still from a video showing a Su-30MKI in the process of releasing a Brahmos cruise missile. (DRDO)

A row of two-seater Rafales: currently one of the primary delivery platforms for Indian nuclear weapons. (IAF)

well as a potential nuclear strike role. Given the finite – and none too generous – strength of the Indian Air Force, it is pertinent to ask if the procurement of specialist aircraft for the nuclear strike and possibly even their escort role are now a necessity.

As to survivability, India has constructed underground hangars in at least three airbases – Adampur, Hindon and Bareilly.[31] However, surface shelters in the form of Hardened Aircraft Shelters (HAS) are not yet available for most of the Su-30MKI fleet. A proposal to build 108 New Generation Hardened Aircraft Shelters with improved protection characteristics is currently pending.[32] HAS are not strictly speaking protection against a pre-emptive nuclear strike but against an airburst they may protect aircraft to some degree. It is also noted that Indian HAS lack blast doors. While this is understandable in order to facilitate the rapid turnaround of aircraft, it does mean that protection is somewhat sub-optimal. Should the Indian Air Force take its nuclear role seriously, then the question of improved HAS must be considered a priority. India's airbases do have one major advantage – munitions storage, fuel supplies and even command and control facilities are all underground. This enhances the survivability of key military assets in the event of a nuclear first-strike.

However, what it does not pay sufficient attention to is the viability of manned aircraft as nuclear strike assets against the intense and well-organised air defences of either China or Pakistan. There is a qualitative difference between delivering a relatively low-yield weapon against an army formation or even a smaller city as compared to delivering one against a hardened, heavily defended counter-force or counter-value target. In this regard, one aspect needs to be considered – air-delivered stand-off munitions with nuclear payloads. While it is as yet too early to speak of an Indian air-launched cruise missile capability, some analysts suggest that the use of nuclear glide-bombs may provide a degree of stand-off capability to attacking aircraft.

Air-Launched Cruise Missiles – Potential Nuclear Role

Besides its ballistic missiles, India has a cruise missile programme which, when completed, would give the country another option for the delivery of nuclear warheads. It must be clarified that India's cruise missile projects have to be divided into two – the Brahmos supersonic system and the Nirbhay subsonic system. While there have been multiple reports of the Brahmos being nuclear capable and/or fitted with a 20 kiloton warhead, it is suggested that this may not be entirely accurate – although the potential certainly exists for such an event.[33] Indeed, the Brahmos is set to experience something of a transformation. Initially claimed to be a 290km range missile, India's accession to the Missile Technology Control Regime enabled the system to be tested to its full range of 450km against both land and sea targets. It is anticipated that a longer-ranged Brahmos – perhaps capable of ranges of up to 900km – will be developed in air, sea and land-launched versions. It is possible that a nuclear warhead could be developed for such an extended range version, although the principal task of the Brahmos would be long-range precision attack.[34]

The Nirbhay, however, is intended to be nuclear capable, although it will also be an important component for convention payload delivery. It is India's first strategic cruise missile with a range exceeding 1,000km with a speed of between 0.8 and 0.9 Mach (speed of sound) with a launch weight of 1,500kg and a length of 6 metres. The payload size of the missile is not yet known but there is speculation that it would be capable of carrying either conventional or nuclear warheads of 24 different types. It is intended to fly at very low altitudes of 20 metres or less to evade radar detection, using a

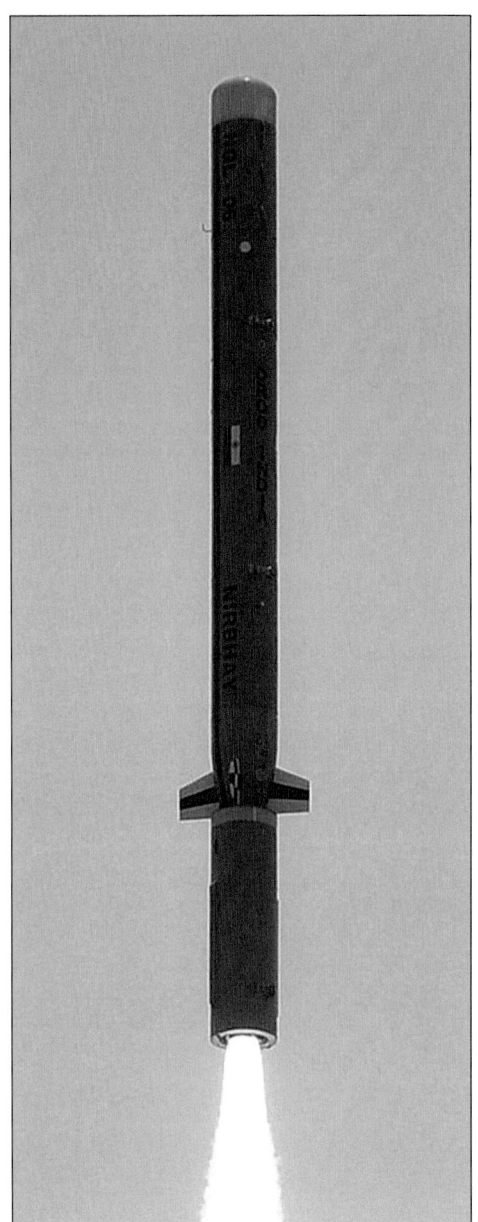

Test launch of the Nirbhay cruise missile: an air-launched variant is undergoing development. (DRDO)

terrain following function which enables it to use mapped contours of the ground for both concealment and for updating its inertial navigation system. Once completed, both systems could be nuclear weapons capable.

The Rest of the Indian Nuclear Arsenal

The strength of the Agni missile force is obviously a closely guarded secret. It has been suggested that the 334th and 335th Missile Groups, which operate the Agni-I and Agni-II missiles respectively, possess 8-12 TELs each. While it has been stated that another group has been raised on the Agni-III, its designation is unknown and whether the TELs are accompanied by vehicles which carry reload missiles is equally unclear. It would be very odd if the total production of Agni missiles still remained in the low single digits per annum. It is in light of this that Dr S. Christopher's comments about 2015 production meeting only 20 percent of requirements is significant.[35]

Though Dr S. Christopher went on to indicate that production of the Agni-I and Agni-II missiles would be increased, the fact that there appears to be a significant disconnect between production and

SFC requirements. As was noted earlier, as early as 1998, directions were given to prepare production for up to 12 Agni missiles per annum.[36]

Quantities of missiles seem to still be a mystery. American assessments suggest in the cases of the Agni-II and Agni-III systems, fewer than ten launchers are operational. This may be on the basis of early reports suggesting that each missile group was to have only eight launchers.[37] This may not be accurate as these numbers do not consider road-mobile variants of the Agni-II and Agni-III nor do the numbers consider missiles and launchers that are kept in storage.

It has been suggested that at least two Agni-II missile trains are operational – perhaps with as many as eight launchers each.[38] Dr Bharat Karnad argues that from the Agni-III onwards, the systems would not be rail-deployed but stored in hardened mountain tunnel complexes. This would make considerable sense to further reduce detection and also to enhance survivability.[39] It might be expected that ready-use and stored missiles would be so accommodated. Given the fact that at least two such complexes are already functioning, the conditions of storage and the maintenance of stored missiles must be examined on a continuous basis.[40]

To date there have been 12 Agni-II tests, more than a dozen Agni-I tests, nine Agni-III tests. eight Agni-IV tests and seven Agni-V tests. As noted earlier, India has had the habit of initiating the deployment of missiles after three successful technical trials. Extrapolating production from these tests is problematic.

If it is assumed that production commenced after the third successful technical trial, and that user trials would not involve testing more than one in five of any production batch, then it might be suggested that production of each type exceeds 20 missiles. This might also tally with the rough estimates available for the number of warheads held by India – if it is speculated that between 20 and two dozen of each of the Agni-I through -IV variants is in service to date, this would give India an eventually deployable strength of perhaps over 200 land-based missiles with some 80 launchers.

To these must be added – albeit as yet in potential – the SSBNs of the Indian Navy. In 2009, the INS Arihant – the lead vessel of a class of at least three ballistic missile submarines (SSBN) was launched. The vessel is still undergoing evaluation and sea trials although reports of emerged its commissioning in August 2016 and is an ambitious attempt to give India an operational and survivable nuclear triad.[41] Many Indian nuclear theorists and strategists have held that a force of SSBNs is essential for India to have a fully survivable nuclear deterrent and the development of the INS Arihant has largely been welcomed as a major step forward.

The INS Arihant (S-2) is a relatively small SSBN, displacing around 6,000 tonnes powered by an 83MW pressurised light-water reactor with enriched uranium fuel. The vessel has a top submerged

The Agni-II was India's first IRBM and the backbone of its nuclear arsenal for years. (DRDO)

An Agni-IV in the process of lift-off from its TEL. (DRDO)

A test launch of an Agni-V – India's most potent ballistic missile. (DRDO)

The K-4 was first tested on 24 March 2014 from a submerged pontoon.[44] Another pontoon launch followed on 7 March 2016.[45] Thereafter it was reportedly tested from the INS Arihant itself on 31 March 2016.[46] This latter claim is probably untrue and refers to an ejection test from the INS Arihant rather than a fill test. It would be surprising if such a major development were not highlighted in some way but in any event, development of the K-4 is clearly well in progress. Four such missiles are to arm the Arihant-class and the K-4 bears a striking resemblance to the land-based Agni-III dimensionally and in respect of its performance.

The K-4 missile took some major steps towards deployment aboard the INS Arihant when, on 19 January 2020, the first submerged launch from a pontoon was conducted to the missile's full range.[47] This was followed by a repeat, limited range test on 24 January 2020.[48] At 12 metres in length, a weight of 17 tonnes and with a 3,500km range, the K-4 is a two-stage solid-fuel missile and in the aftermath of the second trial, it was reported that the missile was ready for induction and fitment aboard the INS Arihant.

Despite reports of commissioning and submerged K-15 and K-4 missile launches, it is suggested that some degree of caution should be exercised when claiming that an Indian sea-based deterrent is operational, although the potential is very evident as the submerged pontoon launches demonstrate.

However, even armed with the modest K-15 missile, the INS Arihant has been commissioned and on 5 November 2018, it was disclosed that the vessel had completed a deterrent patrol of 20 days, ending on 4 November 2018.[49] This, at the very least, indicates that the vessel has some operational capability and is considered to be deployable by the Indian SFC and Indian Navy

It is only when other submarines of the Arihant-class are commissioned will India be truly deemed to have completed its sub-surfaced based nuclear deterrent. Three more such submarines are in various stages of construction. One of these – the INS Arighat (S-3) – has been launched and is undertaking trials and may be commissioned in 2022.

Given that some estimates put the number of Indian nuclear warheads at between 110 and 130, the figure of 200 missiles and 80 launchers plus aircraft-delivered weapons to be included in the overall total of deliverable warheads may give an indication that the estimates of the Indian arsenal are somewhat understated. It is suggested that a realistic estimate of India's missile arsenal, as it emerges in the future, is going to be as listed as shown in Table 8.[50]

When the potential air-delivered weapons of the Indian Air Force are added – whether through free-fall or glide bombs or from air-launched cruise missiles, it is evident that the IAF has a significant nuclear role to play, despite being overshadowed in the Indian nuclear arsenal by the Indian force of ballistic missiles – both mobile land-based systems and, increasingly, SLBMs.

speed of 24 knots. The INS Arihant can carry twelve K-15 missiles or four K-4 missiles.

India is developing two SLBMs – the 750km range K-15 and the 3,500km range K-4 to be deployed aboard the Arihant class of ballistic missile submarines. The K-15 underwent at least 12 development trials from a submerged pontoon aimed at simulating a submarine.[42] However, on 25 November 2015, an unarmed K-15 was purportedly fired from the INS Arihant.[43] It should be stated that this has not yet been confirmed by Indian officials and no photographs have emerged of such a launch from the Arihant.

Table 8: Strategic Forces Command, Indian Army (Estimate for 2021)[51]		
Unit	Base	Notes
334th Missile Group	Secunderabad	Agni-I, 8-16 launchers, 24–48 missiles
335th Missile Group		Agni-II, 8-16 launchers, 24–48 missiles
336th Missile Group		Agni-III, 8-16 launchers, 24–48 missiles
? Missile Group		Agni-IV, 16 launchers, 48 missiles
? Missile Group		Agni-V 16 launchers, 48 missiles
333rd, 444th & 555th Missile Groups		Prithvi SS-150, SS-250 & SS-350 – total of over 200 SRBMs deployed
? Missile Group		Shaurya missiles – very limited deployment

7
INDIA'S MILITARY SPACE EFFORTS: THE IAF'S FINAL FRONTIER

One of the least appreciated aspects of India's aerospace is one which is not specifically intended for military purposes. India has made considerable progress in the use of satellites for military purposes and, while to date, India has not deployed any dedicated space-based early-warning systems to complement its intended land-based missile defence radars, it has made some progress in developing secure communications and surveillance satellites. Now that India has moved to create a Tri-Service military space agency, one must view space as an important domain for the IAF and one which will inevitably grow in importance as the IAF gives shape to the Defence Space Agency.

India's satellite saga began in the 1970s while its development of effective launch vehicles achieved its first success in 1980 but it gained much more traction in the 1990s. India does not have a dedicated space programme, and its use of space technology for military purposes is decidedly secretive and probably makes extensive use of dual-purpose satellites for military functions. While a specialised military space programme is perhaps currently not cost-effective, it is suggested that India needs to make more effective use of its existing assets, with perhaps some accretion in the number of satellites.

At the outset, it should be noted that India's capability of launching satellites into both Polar Synchronous Orbit, through the Polar Satellite Launch Vehicle (PSLV), and Geostationary Orbit, increasingly through various iterations of the Geostationary Satellite Launch Vehicle (GSLV), does offer potential benefits to the Indian missile programme; however, to date, India has not chosen to overtly use technology from its space programme to enhance the capabilities of its missiles. This was, in the past, to ensure that the Indian Space Research Organisation (ISRO) did not fall foul of arms control regimes such as the Missile Technology Control Regime (MTCR) of which India was not then a member. With India now getting membership to the MTCR, the opportunity exists for somewhat greater collaboration. India's space programme's launch vehicles and its ability to launch multiple satellites into orbit offers potential technologies which could be used to enhance the range of India's missiles as well as assist in its development of MIRVs for those missiles. In the past, there has been some speculation that India's SLV systems could be morphed into ballistic missiles. The trajectory of the Indian ballistic missile programme to date suggests that any such concern is unwarranted.

Nonetheless, in respect of its nuclear posture, aside from enhancing its conventional forces through improved communications and surveillance, the opportunity exists for India to explore the use of space technology as has been done by other nuclear powers to enhance its early warning capabilities and to possibly explore the idea of space-based weapons.

India's Efforts

On 10 June 2008 the then Defence Minister A.K. Antony announced the formation of an Integrated Space Cell under the aegis of the Integrated Defence Services Head Quarters which would act as a single entity to facilitate integration among the Armed Forces, the Department of Space and the Indian Space Research Organisation (ISRO), operated jointly by the three service arms, the DRDO and the ISRO. While India has not yet developed a dedicated space command – such as the Russian Space Forces or the United States Space Command – India does make extensive use of space technology for its military needs but has chosen the path of dual-use satellites to fulfil its requirements.1 This was expanded into a Defence Space Agency – a tri-service entity – in 2019.

Indian Satellites with Military Applications

India today has some 14 operational satellites dedicated to remote sensing, making the Indian Remote Sensing (IRS) series the largest civilian remote-sensing constellation in orbit. All of the IRS satellites are placed in polar sun-synchronous orbit, being launched by the Polar Satellite Launch Vehicle (PSLV), making both the satellites and launch vehicle entirely India. The IRS satellites provide data in a variety of spatial, spectral and temporal resolutions, some having a spatial resolution of one metre or below which have definite military applications.

An early experiment in this regard was the Technology Experimental Satellite (TES) which was launched in October 2001. The TES had panchromatic cameras capable of producing images of one metre resolution and was perhaps India's first ostensible reconnaissance satellite. This does not mean that earlier IRS satellites did not have military applications but rather reflects the greater military utility of the lower resolution afforded by the TES.[2]

One of the more recent satellites with clear military applications was the Radar Imaging Satellite 2 (RISAT-2) which has a synthetic aperture radar (SAR) purchased from Israel Aerospace Industries (IAI). It has a day-night, all-weather monitoring capability and has a resolution of one metre, enabling it to track ships at sea. The RISAT-1, launched later than the RISAT-2, was also fitted with a SAR system and augments the former in surveillance tasks.[3]

The CARTOSAT family of satellites – a subset of the IRS family – are probably the most capable of India's military satellites. The CARTOSAT-2A is a dedicated military satellite whose capabilities are as yet unclear, but it is capable of being steered to enable the imaging of any particular area with greater frequency. It reportedly has a resolution of 2.5 metres.

Its predecessor – the CARTOSAT-2 – carries a state-of-the-art panchromatic (PAN) camera that can take black and white photographs in the visible region of the electromagnetic spectrum. The swath covered by these high-resolution PAN cameras is 9.6km and their spatial resolution is 80 centimetres. CARTOSAT-2 can be steered up to 45 degrees along as well as across the track and is capable of providing scene-specific spot imagery. Its successor, CARTOSAT-2B, offers multiple spot scene imagery.

In 2017, the CARTOSAT-2E was launched. This provided scene-specific images with a spatial resolution of less than 60 centimetres and marked a remarkable improvement in India's satellite imagery. The latest member of the CARTOSAT family, the CARTOSAT-2F, was launched in January 2018. It has four MX detectors with bandpass filters between 450 and 860 nanometers which can deliver imagery at a two-metre ground resolution along a 10 kilometre swath. While its principal tasks are for disaster management, cartography, and environmental monitoring, its military applications are evident.[4] One might expect further improvements in follow-on satellites.

The Polar Satellite Launch Vehicle – the backbone of the India's launch vehicle fleet as of the early 2020s. (ISRO)

The IRS P-3 remote sensing satellite. (ISRO)

Aside from remote sensing satellites, ISRO has fielded a wide variety of communications satellites. The INSAT series of satellites has provided transponders for communications for several decades. Of the 24 satellites put into orbit, 11 are currently operational with ISRO making more use of the Indian Geostationary Satellite Launch Vehicle (GSLV), as opposed to the French Ariane rocket, as the reliability of the GSLV improves. INSAT-4F, also known as GSAT-7, is a dedicated military communications satellite with the Indian Navy making extensive use of the satellite to facilitate communications between its ships. GSAT-6, launched in 2015, also purportedly provides secure communications to several strategic end-users in India.

More recently, India has completed all the building-blocks for its own satellite navigation system. Six IRNSS satellites are currently in operation, providing the foundation for the development of an Indian navigation satellite system – NAVIC. This will provide both a standard positioning service for civilian users and an encrypted positioning service for military users.

While six satellites are sufficient to start the NAVIC system, an attempt to place a seventh satellite into orbit failed in 2017. This satellite, which was to replace the earliest member of the family – IRNSS-1A – was to provide the necessary redundancy for the NAVIC system. There are plans, however, to increase the number of satellites to eleven, thus increasing the coverage area.

Indian Military Use of Satellites

India's employment of satellites for military purposes has been gradually revealed. In 2016, half a dozen ISRO satellites were used to obtain ground information for the surgical strike carried out by the Indian Army against terrorist targets in Pakistan-occupied Kashmir (PoK). Prior to this, the Indian Navy used the GSAT-7 to assist in its search and rescue operations and was able to seamlessly

India's Anti-Satellite Missile – the PDV Mk.II – seen on its mobile launcher, pending another test-firing. (DRDO)

network some 60 ships and 75 aircraft during the 2014 Theatre-level Readiness and Operational Exercise conducted in the Bay of Bengal.[5]

Shortly after the 2017 launch of the CARTOSAT-2E, it was revealed that some 13 satellites were being used for surveillance purposes. This number is set to grow as newer versions of the IRS series begin to enter service. However, barring a few examples, none are dedicated military satellites, being used for both civilian and military purposes.[6]

India's National Technical Research Organization (NTRO) which is controlled by the Research and Analysis Wing, India's premier intelligence agency, makes extensive use of IRS, RISAT and CARTOSAT data to aid it building a comprehensive intelligence picture.

ASAT Test

India also has a nascent ASAT capability though it has not been demonstrated to date. In 2017, an Indian exo-atmospheric ballistic missile interceptor, the PDV, achieved an interception at an altitude of 97km. This is not adequate as yet for an ASAT system but demonstrates a potential capability similar to that of the Israeli Arrow-3 missile. The interception of a satellite, however, is different to that of a ballistic missile. The successful intercept achieved by the US Navy using an ABM system points to a potential dual purpose. In India's case, it has also been suggested that a modified Agni-V missile could serve as a viable ASAT weapon with the propulsion system mated to a dedicated kill-vehicle. However, without a dynamic test, India's ASAT capability was at best theoretical.

On 27 March 2019, India announced that it had intercepted a satellite – later identified as an Indian Microsat-R – at an altitude of 300km with a ground-based missile. This, at one stroke, dramatically enhanced India's capabilities in the spheres of both Anti-Satellite defence and ballistic missile defence. The project – codenamed XSV-1 – led to the fielding of a missile known as the PDV (Prithvi Defence Vehicle) Mk.2 which has at least some traits in common with the PDV exo-atmospheric interceptor of the Indian Ballistic Missile Defence Programme.[7]

The PDV Mk.II clearly builds on the PDV and is a three stage missile. While details are as yet unclear, the upper stage clearly derives from the PDV but its lower stages are much more powerful to allow for the increased performance necessary to reach an altitude of 300km.[8]

The PDV Mk.II looks to be an amalgam of PDV kill technology combined with the booster stage of a ballistic missile – probably the submarine launched ballistic missile K4. The missile has a kinetic energy warhead which does not explode but which destroys its target by sheer velocity and mass.

This requires the system to have a hit-to-kill capability as a near miss will not suffice to destroy the target. This was successfully demonstrated during the 29 March intercept and speaks highly of the accuracy and the precision of the intercept.[9]

The intercept target and its altitude of 300km was selected to minimise debris from the intercept. However, the chairman of DRDO G. Sateesh Reddy has indicated that the

CARTOSAT-2C undergoing testing prior to launch into the space. (ISRO)

GSLV – India's largest space launch vehicle. (ISRO)

PDV Mk.II can intercept all satellites up to an altitude of 1,000km.[10] This clearly places India's capabilities close to those demonstrated by China, Russia and the United States.

Shifting to a Deployable System

One test, no matter how successful, does not necessarily make a deployable system. While there is no immediate necessity for another ASAT test, simulations should continue with a view to perfecting a deployable system. Moreover, the PDV Mk.II has definite BMD applications and consideration should be given to developing suitable targets so that India's indigenous exo-atmospheric intercept capability can be perfected.

India is Moving in the Right Direction

The closest India has come to outlining a vision for the use of space technology to enhance its military was the Indian Air Force Defence Space Vision 2020, which outlined ways to harness satellite resources to significantly boost India's defence preparedness. The Defence Space Agency is a step in the right direction, but it is time to look again at whether there is a need for a much larger organisation. Outlined in 2011, the Defence Space Vision focused on military space applications in the spheres of intelligence, reconnaissance, surveillance, communications and navigation. It envisaged the need for several dedicated military satellites as well as the need for a specialised Space Command as tri-service entity and will take aerospace power into an entirely new dimension in capability.

While the Defence Space Agency is now fully operational, without a clear vision, resources and a path forward, it is submitted that such an organisation would be little more than an entity to enable networking and perhaps a certain amount of planning. Will the Defence Space Agency, for example, have control of the satellites to the extent of adjusting orbits where necessary? Furthermore, as all three services have a need for both satellite intelligence and communications, it was a correct move to establish the Defence Space Agency as a tri-service command.

These are fundamentally administrative questions but at this juncture, India needs to ascertain whether it requires a constellation of specialised built-for-purpose military satellites or whether the current dual-use satellites are a more cost-effective way of using resources. This latter point cannot be understated as India will not be able to afford the prodigious duplication of effort that parallel civilian and military satellite programmes would envisage. It is worth remembering that the reconnaissance satellite concept was created and developed at a time when civilian imaging satellites were, at best, in their infancy. Nowadays, civilian satellites can offer imagery rivalling that of dedicated military satellites.

With improved resolutions, the line between civilian and military satellites is being increasingly blurred, especially when it relates to quality imagery. The IRS, CARTOSAT and RISAT dual-use satellites are effectively used by the Indian military. At this point, perhaps concentrating on an improved image resolution, secure data transmission and processing is perhaps the order of the day rather than incurring the expense of dedicated military satellites.

This does not preclude the need for prioritising military imagery requirements, but it does pose the question as to whether the expense of dedicated satellites is needed. This expense could be justified if it was the intention to create a global system of near-constant surveillance along the lines of that developed by the United States. In practical terms, however, it is unlikely that this would be cost-effective for India, though the need for such a constellation of satellites could emerge in the medium-term. It is not possible to transpose the path adopted by the other powers directly onto India's requirements in this regard. For a decision on the need for dedicated military surveillance satellites, India must first define which areas it needs to keep under constant observation and the systems required to do this. If this remains limited to China, Pakistan and the Indian Ocean and Arabian Sea regions, it is not inconceivable that the dual-purpose satellites currently in use could suffice for this purpose and meet India's core military requirements.

With respect to military communications, the needs of the services in terms of transponders must be quantified. Just as the Navy is satisfied with the GSAT-7, the other two services may require similar satellites of their own to ensure secure communications with additional transponders from civilian satellites, with necessary security being used for redundancy. It is unlikely, therefore, that any more than a few dedicated military communications satellites will be needed and even then, the number and type would need to be specified. With respect to navigation, the dual-use IRNSS system is being made operational as soon as practicable and will provide additional capability. The need for dedicated military navigation satellites is not envisaged nor may it be practical especially given the emergence of the IRNSS system as a viable navigational asset.

There may be, however, a growing requirement for dedicated military early-warning satellites as the missile threat from both China and Pakistan grows. Indeed, satellites may offer the only means of launch detection for India. In addition, there may be a requirement for dedicated SIGINT satellites which can bolster intelligence gathering capabilities. In neither of these spheres has India shown any appreciable sense of urgency. This is particularly evident in the failure of India to embrace the urgency of space-based missile warning systems to enhance its ballistic missile early-warning network. It is strange that despite considerable effort in the development of land-based interceptors and radar systems, India has been so reluctant to develop space-based early warning capabilities.

Finally, India is clearly giving consideration to developing a viable and working ASAT capability and demonstrating the same. India has made a very positive step forward with the dynamic testing of PDV-II in the ASAT role, with sufficient success as to demonstrate a definitive capability in that sphere. However, while the system clearly works, India should move towards at least a limited deployment of the system, thus enhancing the IAF's ASAT and BMD capabilities.

In conclusion, India has now developed the desired Defence Space Agency as a fully functional entity as a Tri-Service Space Agency, with military space capabilities that are not inconsiderable. What is lacking is a sense of direction, a clear statement of military space requirements and a closer relationship between the military and space authorities. It is perhaps in this regard that the Defence Space Agency can come into its own and facilitate and foster this relationship with a view eventually morphing into a tri-service Space Command. This will have to interact with the Air Force's existing command structure and the creation of an Aerospace Command is not beyond the realms of possibility.

8
CONCLUSION

After 90 years of existence, the IAF has come a long way in terms of its capability, capacity, infrastructure and doctrine. From flying decrepit Westland Waptis, it is now participating in a tri-service Defence Space Agency with dedicated assets. The force has been tested in battle and shown itself to be adaptable, innovative and ready to learn from its mistakes.

Like many air forces, built-up over the decades with readily available Soviet equipment, the IAF faces significant challenges in improving the quality of its assets while simultaneously retaining and later augmenting its numbers. The next few years will be challenging for the IAF as it upgrades existing platforms and invests in indigenous projects, the fruits of which are now becoming evident. The IAF has invested heavily in indigenous products for its helicopter fleet and its air defences and its faith in the Tejas combat aircraft is growing. However, budgets are never unlimited and the pressure to make do with less than optimal resources is always present. However, the escalating tensions with China may provide the impetus to encourage the indigenisation trend.

As one of the custodians of India's nuclear deterrent, the IAF has shown itself to be ready and willing to adapt its doctrine to new challenges and as integration of the armed services gathers pace in India, the IAF will find itself part of theatre commands, in addition to having responsibility for air defence.

The IAF thus faces its 90th anniversary with a combination of challenges and opportunities. It is at the cusp of completely transforming itself from a moderately modern force to one with cutting edge technologies. Its progress to achieving its full potential will be most interesting to watch.

BIBLIOGRAPHY

Books

Anon., *Basic Doctrine of the Indian Air Force 2012* (Indian Air Force: New Delhi, 2012)

Anon., *Joint Doctrine Indian Armed Forces* (Headquarters, Integrated Defence Staff, New Delhi: 2017)

Aroor, S. & Singh, R., *India's Most Fearless 2* (Gurgaon: Penguin, 2019)

Babbage, R. & Gordon, S. eds, *India's Strategic Future* (London: MacMillan, 1992)

Badri-Maharaj, S., *Kargil 1999 – South Asia's First Post-Nuclear Conflict* (Warwick: Helion, 2020)

Badri-Maharaj, S., *The Armageddon Factor: Nuclear Weapons in the India-Pakistan Context* (New Delhi: Lancer, 2000)

Badri-Maharaj, Sanjay, *Indian Nuclear Strategy – Confronting the Potential Threat from China and Pakistan* (New Delhi: CLAWS, 2019)

Badri-Maharaj, Sanjay, *Nuclear India – From Reluctance to Triad* (Warwick: Helion, 2021)

Bajpai, K., Chari, P., Cheema, S., Cohen S, & Ganguly, S. (eds), *Brasstacks and Beyond: Perception and Management of Crisis in South Asia* (New Delhi: Manohar, 1995)

Baranwal, J. ed., *SP's Military Yearbook 1992–93* (New Delhi: Guide Publications, 1993)

Burrows, W.E., & R. Windrem, *Critical Mass: The Dangerous Race for Superweapons in a Fragmenting World* (New York: Simon & Schuster, 1994)

Chari, P.R., *Indo-Pak Nuclear Standoff: The Role of the United States* (New Delhi: Manohar Publishers, 1995)

Chellaney, B., (ed.), *Securing India's Future in the New Millennium* (New Delhi: Orient Longman Limited, 1999)

Chellaney, B. *Nuclear Proliferation: The US-Indian Conflict* (New Delhi: Orient Longman Ltd, 1993)

Chengappa, Raj, *Weapons of Peace* (New Delhi: Harper Collins Publishers India, 2000)

Chordia, A. K., *Operation Cactus: Anatomy of One of India's Most Daring Military Operations* (New Delhi: Centre for Air Power Studies, 2018)

Conboy, K. & Hannon, P., *Elite Forces of India and Pakistan* (London: Osprey, 1992)

Donald, D. & Lake, J., *Encyclopaedia of World Military Aircraft: Volume Two* (London: Aerospace Publishing Ltd., 1994)

Kanwal, Gurmeet, *Nuclear Defence: Shaping the Arsenal* (New Delhi: Knowledge World, 2001)

Karnad, Bharat, (ed.), *Future Imperilled: India's Security in the 1990s and Beyond* (New Delhi: Viking Penguin India, 1994)

Karnad, Bharat *India's Nuclear Policy* (Westport, Connecticut: Praeger Security International, 2008)

Karnad, Bharat, *Nuclear Weapons and Indian Security* (New Delhi: Macmillan India Ltd, 2002)

Karnad, Bharat, *Why India is not a Great Power (Yet)* (New Delhi: Oxford University Press, 2015)

Kavic, L.J., *India's Quest for Security* (London: University of London Press, 1967)

Kumar, B., *An Incredible War – IAF in Kashmir War 1947–1948* (New Delhi: Centre for Air Power Studies, 2007)

Kumar, B., *Operation Pawan: Role of Airpower with IPKF* (New Delhi: Manohar, 2015)

Kumar, B., *Unknown and Unsung – Indian Air Force in Sino-Indian war of 1962* (New Delhi: Centre for Air Power Studies, 2013)

Menon, Shivshankar, *Choices: Inside the Making of India's Foreign Policy* (Penguin Random House India & Washington DC: Brookings, 2016)

Mirchandani, G., *India's Nuclear Dilemma* (New Delhi: Popular Book Services, 1968)

Nair, V.K., *Nuclear India* (New Delhi: Lancer International, 1992)

Palsokar, R.D, *Minimum Deterrent: India's Nuclear Answer to China* (Bombay: Thacker & Co. Ltd, 1969)

Perkovich, George, *India's Nuclear Bomb: The Impact on Global Proliferation* (London: University of California Press, 1999)

Prasad, S.N., ed., *The India-Pakistan War of 1965 – A History* (New Delhi: Natraj, 2011)

Prasad, S.N., ed., *The India-Pakistan War of 1971 – A History* (New Delhi: Natraj, 2014)

Rikhye, R., Singh, P. & Steinemann, P., *Fiza'Ya: Psyche of the Pakistan Air Force* (New Delhi: S. A. S., 1991)

Rikhye, R., *The War that Never Was* (New Delhi: Chanakya Publications, 1988)

Sethi, Manpreet, *Nuclear Strategy: India's March Towards Credible Deterrence* (New Delhi: KW Publishers, 2009)

Singh, Jasjit, 'A Nuclear Strategy for India', in Jasjit Singh (ed.), *Nuclear India* (New Delhi: Knowledge World in association with Institute for Defence Studies and Analyses, 1998)

Singh, S., *India and the Nuclear Bomb* (New Delhi: S. Chand & Co. Ltd, 1971)

Spector, L., *Nuclear Ambitions: The Spread of Nuclear Weapons 1989–90* (Boulder, Colorado: Westview Press, 1990)

Subramanyam, K. *Indian and the Nuclear Challenge* (New Delhi: Lancer, 1986)

Tiwary, A.K., *Indian Air Force in Wars* (New Delhi: Lancer, 2012)

Articles

'Agni-V can reach targets 8,000 km away: Chinese expert', *The Hindu* (21 April 2012), <http://www.thehindu.com/todays-paper/tp-in-school/agniv-can-reach-targets-8000-km-away-chinese-expert/article3337202.ece>

'Cabinet Committee On Security Reviews Progress In Operationalizing India's Nuclear Doctrine', <http://pib.nic.in/archieve/lreleng/lyr2003/rjan2003/04012003/r040120033.html>

'China, India commence withdrawal of forces from shared border – Chinese Defense Ministry', *TASS* (10 February 2021) <https://tass.com/world/1254813>

'Defence Acquisition Council okays purchase of Russian air defence missile systems', *Press Trust of India* (17 December 2015), <http://www.dnaindia.com/india/report-defence-acquisition-council-okays-purchase-of-russian-air-defence-missile-systems-2156916>

'Draft Indian Nuclear Doctrine', *Arms Control.org*, <https://www.armscontrol.org/act/1999_07-08/ffja99>

'IAF successfully test fires air-to-air 'Astra' missile from Sukhoi-30 MKI' *Hindustan Times* (19 September 2019) <https://www.hindustantimes.com/india-news/iaf-successfully-test-fires-air-to-air-astra-missile-from-sukhoi-30-mki/story-JZ8ijuSchUOfACTCOpbXSM.html>

'India acquires Green Pine radars from Israel', *Press Trust of India* (28 June 2002), <https://timesofindia.indiatimes.

com/india/India-acquires-Green-Pine-radars-from-Israel/articleshow/14351441.cms>

'India Begins User Trials of Russian Air-to-Air Missiles, Destroys UK-Made Drone in Drill' *Sputnik* (20 June 2019) <https://sputniknews.com/20190620/india-begins-user-trials-of-russian-air-to-air-missiles-destroys-uk-made-drone-in-drill-1075979034.html>

'India building nuclear-proof bunkers for top leaders', *Silicon India* (23 September 2003), <http://www.siliconindia.com/shownews/India_building_nuclearproof_bunkers_for_top_leaders-nid-20887-cid-Top.html>

'India capable of building nuke deterrence up to 200 kilotons: Kakodkar', *Deccan Herald* (24 September 2009) <https://www.deccanherald.com/content/27047/india-capable-building-nuke-deterrence.html>

'India conducts first night trial of Agni-II missile', *India Today* (16 November 2019), <https://www.indiatoday.in/india/story/india-conducts-first-night-trial-of-agni-ii-missile-1619704-2019-11-16>

'India has fissile material for 2,000 warheads: Pak media', *Times of India* (10 September 2015), <http://timesofindia.indiatimes.com/world/pakistan/India-has-fissile-material-for-2000-warheads-Pak-media/articleshow/48895568.cms>

'India loses two jets', *BBC News,* (27 May 1999) <http://news.bbc.co.uk/2/hi/south_asia/354120.stm>

'India Plans Next Generation Bombproof Shelter for 108 Fighter Jets' *Sputnik News* (3 July 2017), <https://sputniknews.com/military/201707031055187413-india-shelter-jets>, accessed 4 September 2017

'India successfully test-fires 3,500 km nuclear-capable missile K-4', *New Indian Express* (20 January 2020), <https://www.newindianexpress.com/states/odisha/2020/jan/20/india-successfully-test-fires-3500-km-nuclear-capable-missile-k-4-2091838.html>

'India successfully test-fires supersonic interceptor missile', *NDTV,* (23 November 2012) <https://www.ndtv.com/india-news/india-successfully-test-fires-supersonic-interceptor-missile-505365>

'India successfully test-fires first indigenous 'Rudram' Anti-Radiation Missile from Sukhoi-30' *Financial Express* (9 October 2020) <https://www.financialexpress.com/defence/india-successfully-test-fires-first-indigenous-rudram-anti-radiation-missile-from-sukhoi-30/2101752/>

'India Successfully Test-fires N-capable Agni-I Ballistic Missile', *News 18* (22 November 2016), <http://www.news18.com/news/india/india-successfully-tests-nuclear-capable-agni-i-ballistic-missile-1314397.html>

'India Successfully Test-Fires New Interceptor Missile', *Outlook India* (27 April 2014), <https://web.archive.org/web/20140428163936/http://news.outlookindia.com/items.aspx?artid=838755>

'India successfully test-fires nuclear capable Agni-II missile off Odisha coast', *News 18* (9 November 2014), <http://www.news18.com/news/india/india-successfully-test-fires-nuclear-capable-agni-ii-missile-off-odisha-coast-724952.html>

'India successfully test-fires supersonic interceptor missile', *Tribune India* (15 May 2016), <http://www.tribuneindia.com/news/nation/india-successfully-test-fires-supersonic-interceptor-missile/237107.html>

'India successfully tests nuclear capable Agni IV missile', *The Hindu* (3 December 2014), <http://www.thehindu.com/todays-paper/tp-in-school/india-successfully-tests-nuclear-capable-agni-iv-missile/article6656285.ece>

'India test fires medium range nuclear capable Agni-II missile', *Economic Times* (20 February 2018), <https://economictimes.indiatimes.com/news/defence/india-test-fires-agni-ii-missile-off-odisha-coast/articleshow/62993975.cms>

'India test fires nuclear capable Agni-5 missile, 2nd test in six months', *Economic Times* (10 December 2018), <https://economictimes.indiatimes.com/news/defence/india-test-fires-nuclear-capable-agni-5-missile-2nd-test-in-six-months/articleshow/67023684.cms>

'India test-fires Agni V with range as far as China', *Hindustan Times* (16 September 2013), <http://www.hindustantimes.com/india/india-test-fires-agni-v-with-range-as-far-as-china/story-28IHgfrhxGgUt9XLNPRiwN.html>

'India test-fires nuclear-capable Agni III ballistic missile', *Economic Times* (27 April 2017), <http://economictimes.indiatimes.com/news/defence/india-test-fires-nuclear-capable-agni-iii-ballistic-missile/articleshow/58396268.cms>

'India test-fires nuclear-capable Nirbhay cruise missile', *Times of India* (17 October 2014), <http://timesofindia.indiatimes.com/india/India-test-fires-nuclear-capable-Nirbhay-cruise- missile/articleshow/44845526.cms>

'India to join ICBM club soon – Interview with Dr. S. Christopher', *NDTV* (11 July 2015), <http://www.ndtv.com/video/news/news/india-to-join-icbm-club-soon-374683>

'India to test fire Agni-V by year-end', *The Hindu* (3 June 2011), <http://www.thehindu.com/sci-tech/science/India-to-test-fire-Agni-V-by-year-end/article13821309.ece>

'India's Nuclear Tests: News reports and "eyewitness" accounts', <https://seismo.berkeley.edu/~rallen/research/nuke/India.May98/news.reports.html>

'India's nuclear triad is complete with INS Arihant ending its first deterrence patrol', *The Hindu* (5 November 2018), <https://www.thehindu.com/news/national/ins-arihant-completes-deterrence-patrol-india-declares-nuclear-traid-operational/article25425436.ece>

'Indian army may soon get bio-chem suits', *Rediff.com* (11 May 2011), <http://www.rediff.com/news/report/indian-army-may-soon-get-bio-chem-suits/20110511.htm>

'Indian Navy successfully test fires Dhanush missile: All you need to know', *India Today* (26 November 2015), <http://indiatoday.intoday.in/education/story/dhanush/1/531950.html>

'Interceptor missile scores "direct hit"', *The Hindu* (7 December 2007), <http://www.hindu.com/2007/12/07/stories/2007120761241800.htm>

'Interceptor missile test fired successfully', *NDTV.com* (6 March 2011) <https://www.ndtv.com/india-news/interceptor-missile-test-fired-successfully-449178>

'Long range strategic missile Agni-IV test-fired', *The Hindu* (19 September 2012), <http://www.thehindu.com/sci-tech/science/long-range-strategic-missile-agniiv-testfired/article3914340.ece>

'Massive cover-up followed Balakot airstrike: Italian journalist to WION', *WION* (8 May 2019) <https://www.wionews.com/south-asia/130-170-terrorists-died-following-balakot-airstrikes-says-award-winning-italian-journalist-217030>

'Missile defence shield to be ready in three years: India', *Dawn* (13 December 2007), <https://www.dawn.com/news/280120>

'Nirbhay, India's Indigenous Cruise Missile, Fails Test Midway', *NDTV* (16 October 2015), <http://www.ndtv.com/india-

news/nirbhay-indias-indigenous-cruise-missile-fails-midway-1233086>
'Now, India has a nuclear triad', *The Hindu* (18 October 2016), <http://www.thehindu.com/news/national/Now-India-has-a-nuclear-triad/article16074127.ece>
'NUCLEAR ANXIETY; Indian's Letter to Clinton On the Nuclear Testing'. *New York Times* (13 May 1998), <https://www.nytimes.com/1998/05/13/world/nuclear-anxiety-indian-s-letter-to-clinton-on-the-nuclear-testing.html>
'Nuclear Capable Agni-IV missile successfully test fired', *Economic Times* (23 December 2018), <https://economictimes.indiatimes.com/news/defence/nuclear-capable-agni-iv-missile-successfully-test-fired/articleshow/67215177.cms>
'Odisha: Nuclear capable Agni-II missile successfully test fired', *IBN Live* (7 April 2013), <https://archive.is/20130628175643/> <http://www.bharat-rakshak.com/NEWS/newsrf.php>
'Operation Shakti', *Nuclear Weapons Archive – India's Nuclear Weapons Program*, <https://nuclearweaponarchive.org/India/IndiaShakti.html>
'Pakistan was to deploy nukes against India during Kargil war', <https://m.economictimes.com/news/defence/pakistan-was-to-deploy-nukes-against-india-during-kargil-war/articleshow/50019153.cms>
'Pokhran-II tests were fully successful; given India capability to build nuclear deterrence: Dr. Kakodkar and Dr. Chidambaram, *Press Information Bureau* (24 September 2009), <http://pib.nic.in/newsite/PrintRelease.aspx?relid=52813>
'Press Statement by Dr. Anil Kakodkar and Dr. R. Chidambaram on Pokhran-II tests', *Press Information Bureau* (24 September 2009), <http://pib.nic.in/newsite/PrintRelease.aspx?relid=52814>, accessed 20 September 2017
'Prithvi missiles to be replaced by more-capable Prahar: DRDO', *The Hindu Business Line* (30 June 2013), <http://www.thehindubusinessline.com/news/prithvi-missiles-to-be-replaced-by-morecapable-prahar-drdo/article4866081.ece>
'Russia hands over commercial offer of 21 MiG-29 fighters to India' *ANI* (21 July 2021) <https://www.aninews.in/news/world/asia/russia-hands-over-commercial-offer-of-21-mig-29-fighters-to-india20210721060210>
'Shadow of an Indian H-Bomb', *Foreign Report* (13 December 1984), p.1.
'Smiling Buddha', *Nuclear Weapons Archive*, <http://nuclearweaponarchive.org/India/IndiaSmiling.html>
'South Asian Nuclear-Weapon-Free Zone', *The Arms Control Reporter* (1993), p.454. B.176.
'Sub-sonic cruise missile "Nirbhay" successfully test-fired', *The Hindu* (16 April 2019), <https://www.thehindu.com/news/national/other-states/sub-sonic-cruise-missile-nirbhay-successfully-test-fired/article26849032.ece>
'The Long Pause 1974–1989', *Nuclear Weapon archive*, <http://nuclearweaponarchive.org/India/IndiaPause.html>
'The Mirage 2000 in Kargil', <http://www.bharat-rakshak.com/IAF/history/kargil/1056-pcamp.html>
Anantha Krishna M., 'India's Medium Weight Fighter set to fly into detail design phase', *Onmanorama* (3 February 2020) <https://www.onmanorama.com/news/india/2020/02/03/india-defence-expo-medium-weight-fighter.html>
Aroor, S., '80% bombs hit target: IAF gives satellite images to govt as proof of Balakot airstrike', *India Today* (6 March 2019) <https://www.indiatoday.in/india/story/indian-air-force-balakot-airstrike-dossier-satellite-images-bombs-1471355-2019-03-06>
Aroor, Shiv, '10 Reasons Why the Indian Rafale is Evolution Itself', *Daily O* (4 July 2017), <http://www.dailyo.in/variety/rafale-aircraft-brahmos-nuclear-defence/story/1/18157.html>, accessed 4 September 2017
Aroor, Shiv, 'India deploys long-range missile Nirbhay to counter Chinese threat at LAC', *India Today* (28 September 2020), <https://www.indiatoday.in/india/story/india-deploys-long-range-missile-nirbhay-to-counter-chinese-threat-at-lac-1726255-2020-09-28>
Aroor, Shiv, 'True BrahMos Unleashed Today, Next 900km Weapon', *Livefist Defence*, <https://www.livefistdefence.com/2017/03/true-brahmos-unleashed-today-next-1000-km-weapon.html>
Badri-Maharaj, S. 'Flight Training in the Indian Air Force', *Geopolitics* (October 2019)
Badri-Maharaj, S. 'Indian Air Force: Capabilities, Capability Gaps and Options', *Geopolitics* (February 2021)
Badri-Maharaj, S. 'It's Not Just About Combat Aircraft', *Geopolitics* (October 2019)
Badri-Maharaj, S. 'New Helicopters – Indigenous Progress', *Geopolitics* (April 2020)
Badri-Maharaj, S., 'India's Military Special Forces', *Geopolitics* (November 2019)
Badri-Maharaj, S., 'Civil Defence Capabilities of the Indian State', *Bharat Rakshak Monitor*, Vol. 4, No. 2, (September–October 2001)
Badri-Maharaj, S., 'The IAF Fleet Needs Rejuvenating, And Needs It Now', *Swarajya* (9 January 2017), <https://swarajyamag.com/defence/the-iaf-fleet-needs-rejuvenating-and-needs-it-now>
Badri-Maharaj, S., 'The Importance of Passive and Active CBRN Defensive Measures', *Institute for Defence Studies and Analyses: IDSA Issue Brief* (17 October 2016), <https://idsa.in/issuebrief/importance-of-passive-and-active-cbrn-defensive-measures_sbmaharaj_171016>
Badri-Maharaj, S., 'The Indian Air Force's Declining Squadron Strength – Options and Challenges', *Institute for Defence Studies and Analyses – Issue Brief* (3 November 2017), <https://idsa.in/issuebrief/the-indian-air-force-declining-squadron-strength_sbmaharaj_031117>
Balachandran, G. & Patil, K., 'Revisiting India's Nuclear Doctrine', *Institute for Defence Studies and Analyses* (27 March 2017), <https://idsa.in/idsacomments/revisiting-india-nuclear-doctrine_gbala-kpatil_270317>
Barnaby, Frank, 'Trials provide data for range of weapon yields', *Jane's Defence Weekly* (27 May 1998)
Basu, Nayanima, 'Armed Forces Undergoing Nuclear, Biological Warfare Training', *The Hindu Business Line* (11 August 2017), <http://www.thehindubusinessline.com/news/national/armed-forces-undergoing-nuclear-biological-warfare training/article9813830.ece>
Bhatia, Vimal, 'IAF to Deploy New Radar near Border Areas', *Times of India* (2 October 2011), <https://timesofindia.indiatimes.com/city/jaipur/IAF-to-Deploy-new-radar-near-border-areas/articleshow/10204373.cms>
Bikhchandani, R., 'Heron, Searcher, Sea Guardian, SWITCH — the many UAVs that make up India's drone arsenal', *The Print* (6 August, 2021) <https://theprint.in/defence/heron-searcher-sea-guardian-switch-the-many-uavs-that-make-up-indias-drone-arsenal/709670/>

Chopra, Air Marshal Anil, 'India's Military Space Program', *South Asia Defence and Strategic Review*, <http://www.defstrat.com/india%E2%80%99s-military-space-program>

DDR Staff, 'DRDO's AAD Ballistic Missile Defence Interceptor Heads Toward Induction With Latest Test', *Delhi Defence Review* (29 December 2017), <http://www.delhidefencereview.com/2017/12/29/drdos-aad-ballistic-missile-defence-interceptor-heads-toward-induction-with-latest-test/>

Dutta, A.N., 'IAF's Balakot strikes killed 130-170 Jaish terrorists, claims Italian journalist', *The Print* (8 May 2019), <https://theprint.in/defence/iafs-balakot-strikes-killed-130-170-jaish-terrorists-claims-italian-journalist/232809/>

Ghoshal, Arkadev, 'India to Procure Anti-Chemical, Anti-Nuclear Suits from US for Rs 480 crore', *IB Times* (12 May 2017), <http://www.ibtimes.co.in/india-procure-anti-chemical-anti-nuclear-suits-us-rs-480-crore-726410>, accessed 23 October 2017

Gupta, S., 'Maritime, air defence theatre commands to be announced by June 2021', *Hindustan Times* 23 March 2021 <https://www.hindustantimes.com/india-news/maritime-air-defence-theatre-commands-to-be-announced-by-june-2021-101616478448044.html>

Gupta, Shishir 'Govt okays induction of nuke-capable Shaurya missile amid Ladakh standoff', *Hindustan Times* (6 October 2020), <https://www.hindustantimes.com/india-news/shaurya-missile-to-be-inducted-in-strategic-arsenal-agni-5-s-sea-version-by-2022/story-bS1100SkwoGLEXW5ANFQuO.html>

Jha, Saurav, 'DRDO's AAD Endo-atmospheric Ballistic Missile Interceptor Hits Bullseye', *Delhi Defence Review* (1 March 2017), <http://www.delhidefencereview.com/2017/03/01/drdos-aad-endo-atmospheric-ballistic-missile-interceptor-hits-bullseye>

Jha, Saurav, 'Hit-to-Kill Successfully Demonstrated By DRDO's PDV Interceptor', *Delhi Defence Review* (25 February 2017), <http://www.delhidefencereview.com/2017/02/25/hit-kill-successfully-demonstrated-drdos-pdv-interceptor/>

Kampani, G., 'New Delhi's Long Nuclear Journey', *International Security*, Vol. 38, No. 4 (Spring 2014)

Kampani, G., 'India's Evolving Civil-Military Institutions in an Operational Nuclear Context', *Carnegie Endowment for International Peace Regional Insight* (30 June 2016), <https://carnegieendowment.org/2016/06/30/india-s-evolving-civil-military-institutions-in-operational-nuclear-context-pub-63910>

Kaushik, K. "Tejas done, focus on three other fighter jets: two for IAF, one Navy" *Indian Express* (5 February 2021) <https://indianexpress.com/article/india/aero-india-2021-tejas-done-focus-on-three-other-fighter-jets-two-for-iaf-one-navy-7175125/>

Khan, J., Kumar, A., Pathak, S., 'Balakot Tapes Expose: 4 Pakistani soldiers died in air strikes, confirm locals', *India Today* (11 March 2019) <https://www.indiatoday.in/india/story/balakot-tapes-expose-pakistan-army-deaths-1475497-2019-03-11>

Krishna, Ananatha M., 'K-15 SLBM is a Beast with Gen-Next Tech', *New Indian Express* (30 January 2013), <http://www.newindianexpress.com/nation/2013/jan/30/k-15-slbm-is-a-beast-with-gen-next-tech-445756.html>

Kumar, Chethan, 'Six new Akash squadrons to give IAF missile muscle', *Times of India* (17 February 2015), <https://timesofindia.indiatimes.com/india/Six-new-Akash-squadrons-to-give-IAF-missile-muscle/articleshow/46269673.cms>

Mallikarjun, Y. & Subramanian, T., 'Agni-V's Maiden Canister Trial a Roaring Success', *The Hindu* (31 January 2015), http://www.thehindu.com/news/national/maiden-canister-trial-of-agniv-a-roaring-success/article6841942.ece

Mallikarjun, Y. & Subramanian, T., 'Nirbhay Strays from Flight Path, Aborted', *The Hindu* (12 March 2013), <http://www.thehindu.com/news/national/nirbhay-strays-from-flight-path-aborted/article4500527.ece>

Mallikarjun, Y., 'Agni-II Missile Test-Fired Successfully', *The Hindu* (17 May 2010), <http://www.thehindu.com/news/national/Agni-II-missile-test-fired-successfully/article16302660.ece>, accessed 5 September 2017

Mallikarjun, Y., 'Agni-III Test-Fired Successfully', *The Hindu* (21 September 2012), http://www.thehindu.com/news/national/agniiii-testfired-successfully/article3922230.ece

Mallikarjun, Y., 'Interceptor Missile Test Off Odisha Coast Fails', *The Hindu* (6 April 2015), <http://www.thehindu.com/news/national/intercepter-missile-test-odisha-coast-wheeler-islanddrdo/article7073662.ece?ref=sliderNews>, accessed 8 November 2017

Mallikarjun, Y., 'Upgraded Interceptor Missile Successfully Hits Virtual Target', *The Hindu* (22 November 2015), <http://www.thehindu.com/news/cities/Hyderabad/upgraded-interceptor-missile-successfully-engages-electronically-simulated-target-missile/article7905298.ece>

Mallikarjun, Y., AMP & Subramanian, T., 'Agni-V Propels India into Elite ICBM Club', *The Hindu* (19 April 2012), <http://www.thehindu.com/news/national/agniv-propels-india-into-elite-icbm-club/article3330921.ece?homepage=true>

Menon, Air Marshal Narayan, 'Ballistic Missile Defence System for India', *Indian Defence Review*, vol. 27 (July–September 2012), <http://www.indiandefencereview.com/spotlights/ballistic-missile-defence-system-for-india/>

Menon, Raja, 'The Nuclear Doctrine: Yoking a Horse and Camel Together', *Times of India* (26 August 1999)

Nagal, B.S., 'Guest Column: Nuclear No First Use Policy: A Time for Appraisal', *Force* (December 2014), <http://forceindia.net/guest-column/guest-column-b-s-nagal/nuclear-no-first-use-policy/>

Narang, Vipin, 'Plenary: Beyond the Nuclear Threshold: Causes and Consequences of First Use', *Carnegie International Nuclear Policy Conference*, Washington DC (20 March 2017), <https://fbfy83yid9j1dqsev3zq0w8n-wpengine.netdna-ssl.com/wp-content/uploads/2013/08/Vipin-Narang-Remarks-Carnegie-Nukefest-2017.pdf>

Nirbhay, India's Indigenous Cruise Missile, Fails Test Midway', *NDTV* (16 October 2015), <http://www.ndtv.com/india-news/nirbhay-indias-indigenous-cruise-missile-fails-midway-1233086>

Panda, Ankit, 'India Inches Closer to Credible Nuclear Triad With K-4 SLBM Test', *The Diplomat* (13 May 2014), <http://thediplomat.com/2014/05/india-inches-closer-to-credible-nuclear-triad-with-k-4-slbm-test/>

Pandit, Rajat, 'After Agni-V Launch, DRDO's New Target is Anti-Satellite Weapons', *Times of India* (21 April 2012), <https://timesofindia.indiatimes.com/india/After-Agni-V-launch-DRDOs-new-target-is-anti-satellite-weapons/articleshow/12763074.cms>

Pandit, Rajat, 'Agni-V with China in Range Tested; Next in Line is Agni-VI, with Multiple Warheads', *Times of India* (27 December 2016), <http://timesofindia.indiatimes.com/india/agni-v-with-china-in-range-tested-next-in-line-is-agni-vi-with-multiple-warheads/articleshow/56191362.cms>

Pandit, Rajat, 'Arihant's N-capable missile "ready to roll"', *Times of India* (25 January 2020), <https://timesofindia.indiatimes.com/india/india-successfully-test-fires-k-4-submarine-launched-missile/articleshow/73589861.cms>

Pandit, Rajat, 'India test-fires nuclear-capable ICBM Agni-V', *Times of India* (18 January 2018), <https://timesofindia.indiatimes.com/india/india-test-fires-nuclear-capable-icbm-agni-v/articleshow/62550347.cms>

Pandit, Rajat. 'Ballistic missile Agni-IV test-fired as part of user trial', *Times of India* (9 November 2015), <http://timesofindia.indiatimes.com/india/Ballistic-missile-Agni-IV-test-fired-as-part-of-user-trial/articleshow/49720522.cms>

Pereira, Violet, 'N-capable Arihant submarine successfully test-fires unarmed missile', *Magalorean* (26 November 2015), <http://www.mangalorean.com/n-capable-arihant-submarine-successfully-test-fires-unarmed-missile/>

Peri, D., 'CCS okays 83 LCAs worth around ₹47,000 cr. for IAF', *The Hindu* (13 January 2021) <https://www.thehindu.com/news/national/ccs-approves-83-lca-mk-1a-jets-worth-over-48000-crore/article33568729.ece>

Peri, Dinakar, 'Now, India has a nuclear triad', *The Hindu* (18 October 2016), <http://www.thehindu.com/news/national/Now-India-has-a-nuclear-triad/article16074127.ece>

Philip, S.A., 'Why India is set to miss 2021 deadline to upgrade Mirage 2000 fighters', *The Print* (7 October 2021) <https://theprint.in/defence/why-india-is-set-to-miss-2021-deadline-to-upgrade-mirage-2000-fighters/746444/>

Philip, Snehesh Alex, 'India's ballistic missile shield ready, IAF & DRDO to seek govt nod to protect Delhi', *The Print* (8 January 2020), <https://theprint.in/defence/indias-ballistic-missile-shield-ready-iaf-drdo-to-seek-govt-nod-to-protect-delhi/345853/>

Prakash, Arun, 'Strategic Policy Making and the Indian System', *Maritime Affairs*, Vol. 5, No. 2 (Winter 2009), pp.22–31

Press Information Bureau, 'Pokhran-II Tests were Fully Successful; Given India Capability to build Nuclear Deterrence: Dr. Kakodkar and Dr. Chidambaram' (24 September 2009), <http://pib.nic.in/newsite/PrintRelease.aspx?relid=52813>

Press Information Bureau, 'Press Statement by Dr. Anil Kakodkar and Dr. R. Chidambaram on Pokhran-II tests' (24 September 2009), <http://pib.nic.in/newsite/PrintRelease.aspx?relid=52814>, accessed 20 September 2017

PTI, 'Agni III could have 5,000km Range: Russian General', *The Hindu*, <http://www.defencetalk.com/forums/missiles-wmds/indian-nuclear-missile-development-news-discussions-7241-11/>

Pubby, M. 'Government approves $400-million plan to procure armed Heron TP drones from Israel', *Economic Times* (14 July 2018) <https://economictimes.indiatimes.com/news/defence/government-approves-400-million-plan-to-procure-armed-heron-tp-drones-from-israel/articleshow/48906195.cms>

Raghuvanshi, V., 'Indian Excludes Foreign Vendors From Its Air-Defense Upgrade', *Defense News* (13 June 2016), <https://www.defensenews.com/global/asia-pacific/2016/06/13/indian-excludes-foreign-vendors-from-its-air-defense-upgrade/>

Rajagopalan, Rajeswari Pillai, 'India's Nuclear Security: Strengths and Gaps', *Observer Research Foundation* (14 June 2017), <http://www.orfonline.org/research/india-nuclear-security-strengths-gaps/>

Rout, H. K., 'EXPRESS EXCLUSIVE: Maiden Test of Undersea K-4 Missile From Arihant Submarine', <http://www.newindianexpress.com/nation/2016/apr/09/EXPRESS-EXCLUSIVE-Maiden-Test-of-Undersea-K-4-Missile-From-Arihant-Submarine-921990.html>

Rout, H. K., 'India successfully test fires Agni-V missile for a reduced range', *New Indian Express* (26 December 2016), <http://www.newindianexpress.com/nation/2016/dec/26/india-successfully-test-fires-agni-v-missile-for-a-reduced-range-1553219.html>

Rout, H.K., 'India to conduct first user trial of Agni-V missile', *New Indian Express* (13 September 2021), <https://www.newindianexpress.com/states/odisha/2021/sep/13/india-to-conduct-first-user-trial-of-agni-v-missile-2357942.html>

Rout, H.K., 'IAF test fires air-to-air missiles ahead of Rafale integration', *New Indian Express* (19 August 2020) <https://www.newindianexpress.com/states/odisha/2020/aug/19/iaf-test-fires-air-to-air-missiles-ahead-of-rafale-integration-2185422.html>

Rout, Hemant Kumar, 'India test fires new generation nuclear capable Agni-Prime missile off Odisha coast', *New Indian Express* (28 June 2021), <https://www.newindianexpress.com/nation/2021/jun/28/india-test-fires-new-generation-nuclear-capable-agni-prime-missile-off-odisha-coast-2322550.html>

Roy, I., 'All You Need To Know About The PDV MK-II: India's Satellite Killer', *Delhi Defence Review* (3 April 2019) <https://delhidefencereview.com/2019/04/03/all-you-need-to-know-about-the-pdv-mk-ii-indias-satellite-killer/>

Saran, Shyam, 'Is India's Deterrent Credible?', *India Habitat Centre* (24 April 2013), <http://www.armscontrolwonk.com/files/2013/05/Final-Is-Indias-Nuclear-Deterrent-Credible-rev1-2-1-3.pdf>

Saran, Shyam, 'Is India's Nuclear Deterrent Credible?', Speech delivered at India Habitat Centre, New Delhi (24 April 2013)

Sawhney, Pravin, 'Decks Cleared For the Contract Signing of S-400 ADMS in December', *Force India* (17 October 2017), <http://forceindia.net/decks-cleared-contract-signing-s-400-adms-december/>

Sengupta, Prasun, 'Arudhra MPR Is EL/M-2084 MMR: Seeing Is Believing', *Trishul-Trident* (4 June 2011), <http://trishul-trident.blogspot.com/2011/06/arudhra-mpr-is-elm-2084-mmr-seeing-is.html>

Sengupta, Prasun, 'IAF's Multi-Phase IACCCS Being Enhanced', *Trishul-Trident* (22 January 2012), <http://trishul-trident.blogspot.com/2012/01/iafs-multi-phase-iacccs-being-enhanced.html>

Sengupta, Prasun, 'Poised for A Hattrick', *Force India* (October 2017), <http://forceindia.net/cover-story/poised-for-a-hattrick/>

Sharma, S., 'MiG-29 Upgraded: IAF Shares MiG-29 Upgrade Pictures From Deployment In Ladakh; Watch Them Soar', *Republic World* (8 April 2021) <https://www.republicworld.com/india-news/general-news/iaf-shares-mig-29-upgrade-pictures-from-deployment-in-ladakh-watch-them-soar.html>

Shukla, Ajai, 'Army Orders Surface to Air Missile, Making it the First Tri-Service Weapon', *Business Standard* (26 September 2017), <http://www.business-standard.com/article/economy-policy/army-orders-surface-to-air-missile-making-it-the-first-tri-service-weapon-117092500988_1.html>

Shukla, Ajai, 'India Launches 5,000-km Range Agni-5 Missile Successfully', *Business Standard* (20 April 2012), <http://www.business-standard.com/article/economy-policy/india-launches-5-000-km-range-agni-5-missile-successfully-112042002020_1.html>

Shukla, Ajai, 'New-Age Agni to boost Pak-Focused Nuclear Deterrent', *Business Standard* (17 December 2016), <http://www.business-standard.com/article/economy-policy/new-age-agni-to-boost-pak-focused-nuclear-deterrent-116121601111_1.html>

Siddiqui, H., 'With upgraded avionics, Jaguar still lacks the thrust', *Financial Express* 29 September 2019 <https://www.financialexpress.com/defence/with-upgraded-avionics-jaguar-still-lacks-the-thrust/1721036/>

Simha, R.K., 'How the IAF Dominated the Skies during Kargil War', *Russia Beyond the* Headlines (26 July 2016) <https://www.rbth.com/blogs/stranger_than_fiction/2016/07/26/how-the-iaf-dominated-the-skies-during-kargil-war_615175>

Singh, Surendra, 'Military using 13 Satellites to Keep Eye on Foes', *Times of India* (26 June 2017), <https://timesofindia.indiatimes.com/india/military-using-13-satellites-to-keep-eye-on-foes/articleshow/59314610.cms>

Somasekhar, M. 'India's "Interceptor" May Make Ballistic Missile Shield Real', *The Hindu Business Line* (12 February 2017), <http://www.thehindubusinessline.com/news/indias-interceptor-may-make-ballistic-missile-shield-real/article9537665.ece>

Subramanian T. & Mallikarjun, Y., 'Agni-II Soars in Success', *The Hindu* (30 September 2011), <http://www.thehindu.com/news/national/agniii-soars-in-success/article2499781.ece>

Subramanian, T. & Mallikarjun, Y., 'Agni-III test-fired successfully', *The Hindu* (8 May 2008), <http://www.thehindu.com/todays-paper/Agni-III-test-fired-successfully/article15218480.ece>

Subramanian, T. & Mallikarjun, Y., 'India successfully Test-Fires Shourya Missile', *The Hindu* (24 September 2011), <http://www.thehindu.com/sci-tech/science/india-successfully-testfires-shourya-missile/article2482010.ece>

Subramanian, T., 'Agni-IV Missile Successfully Test Fired', *The Hindu* (20 January 2014), <http://www.thehindu.com/news/national/agniiv-missile-successfully-test-fired/article5596563.ece>

Subramanian, T.S., 'Interceptor Missile Mission a "Failure"', *The Hindu* (23 May 2016), <http://www.thehindu.com/news/national/Interceptor-missile-mission-a-%E2%80%98failure%E2%80%99/article14333898.ece>

Subramanian, T.S., 'Nirbhay Likely to be Test-Fired in April', *The Hindu* (7 March 2012), <http://www.thehindu.com/todays-paper/tp-national/nirbhay-likely-to-be-testfired-in-april/article2968219.ece>

Subramanian, T.S., 'Nirbhay Missile Test "an utter failure"', *The Hindu* (21 December 2016), <http://www.thehindu.com/news/national/Nirbhay-missile-test-%E2%80%9Can-utter-failure%E2%80%9D/article16915750.ece>

Subramanian, T.S., 'Smashing Hit', *Frontline* (22 December 2007–4 January 2008), <http://www.frontline.in/static/html/fl2425/stories/20080104242512300.htm>

Subramanian,T.S., 'Agni-IV missile successfully test fired', *The Hindu* (2 January 2017), <http://www.thehindu.com/news/national/Agni-IV-missile-successfully-test-fired/article16977450.ece>

Tufail, M. K., 'Role of the Pakistan Air Force During the Kargil Conflict', *CLAWS Journal (*Summer 2009)

Tur, Jatinder Kaur, 'India Developing E-bomb to Paralyze Networks', *Times of India* (29 August 2013), <https://timesofindia.indiatimes.com/india/India-developing-E-bomb-to-paralyze-networks/articleshow/22127411.cms>

Unnithan, Sandeep, 'From India Today magazine: A Peek into India's Top Secret and Costliest Defence Project, Nuclear Submarines', *India Today* (7 December 2017), <http://indiatoday.intoday.in/story/india-ballistic-missile-submarine-k-6-submarine-launched-drdo/1/1104982.html>

Unnithan, Sandeep, 'India has all the building blocks for an anti-satellite capability', *India Today* (27 April 2012), <http://indiatoday.intoday.in/story/agni-v-drdo-chief-dr-vijay-kumar-saraswat-interview/1/186248.html>

Vasani, Harsh, 'India's Anti-Satellite Weapons', *The Diplomat* (14 June 2016), <https://thediplomat.com/2016/06/indias-anti-satellite-weapons/>

Whitfield, Elizabeth, 'Fuzzy Math on Indian Nuclear Weapons', *Bulletin of the Atomic Scientists* (19 April 2017), <http://thebulletin.org/fuzzy-math-indian-nuclear-weapons9343>

Websites

Chopra, S. & Pillarisetti, J., 'Indian Air Force Wings', *Bharat-Rakshak.com,* <http://www.bharat-rakshak.com/IAF/units/bases/282-wings.html#gsc.tab=0>

Chopra, S. & Pillarisetti, J., 'Forward Base Support Units', *Bharat-Rakshak.com,* <http://www.bharat-rakshak.com/IAF/units/bases/284-fbsus.html#gsc.tab=0>

Chopra, S. & Pillarisetti, J., 'Air Force Stations', *Bharat-Rakshak.com,* <http://www.bharat-rakshak.com/IAF/units/bases/283-air-force-stations.html#gsc.tab=0>

Chopra, S. & Pillarisetti, J., 'Indian Air Force Fleet Strength', *Bharat-Rakshak.com,* <http://www.bharat-rakshak.com/IAF/units/others/281-fleet.html>

Chopra, S. & Pillarisetti, J., Unmanned Aerial Vehicle Squadrons', *Bharat-Rakshak.com,* <http://www.bharat-rakshak.com/IAF/units/others/288-uav-squadrons.html#gsc.tab=0>

NOTES

Chapter 1

1. Bharat Kumar, *An Incredible War – IAF in Kashmir War 1947–1948* (New Delhi: Centre for Air Power Studies, 2007), p.267.
2. Bharat Kumar, *Unknown and Unsung – Indian Air Force in Sino-Indian war of 1962* (New Delhi: Centre for Air Power Studies, 2013), pp.272–273.
3. S.N. Prasad, ed., *The India-Pakistan War of 1965 – A History* (New Delhi: Natraj, 2011), pp.239–267.
4. S.N. Prasad, ed., *The India-Pakistan War of 1971 – A History* (New Delhi: Natraj, 2014), pp.205–209 and pp.351–370.
5. Ravi Rikhye, *The War that Never Was* (New Delhi: Chanakya Publications, 1988) pp.137–141.
6. Ravi Rikhye, *The War that Never Was* (New Delhi: Chanakya Publications, 1988) pp.137–141.
7. Rikhye, *The War that Never Was*, pp.137–141.
8. Rikhye, *The War that Never Was*, pp.137–141.
9. Sanjay Badri-Maharaj, *The Armageddon Factor: Nuclear Weapons in the India-Pakistan Context* (New Delhi, Lancer: 2000), pp.205–206.
10. Bharat Kumar, *Operation Pawan: Role of Airpower with IPKF* (New Delhi, Manohar: 2015), pp.173, 201.
11. Kumar, *Operation Pawan: Role of Airpower with IPKF*, pp.173, 201.
12. Kumar, *Operation Pawan: Role of Airpower with IPKF*, pp.173, 201.
13. Rikhye, *The War that Never Was*, pp.137–141.
14. Rikhye, *The War that Never Was*, pp.137–141.
15. Rikhye, *The War that Never Was*, pp.137–141.
16. Ken Conboy & Paul Hannon, *Elite Forces of India and Pakistan* (London, Osprey: 1992), pp.15–19; Sanjay Badri-Maharaj, 'India's Military Special Forces', *Geopolitics* (November 2019), pp.18–24.
17. Badri-Maharaj, 'India's Military Special Forces', pp.18–24.
18. Badri-Maharaj, 'India's Military Special Forces', pp.18–24.
19. Kumar, *Operation Pawan: Role of Airpower with IPKF*, p.395.
20. Ashok K. Chordia, *Operation Cactus: Anatomy of One of India's Most Daring Military Operations* (New Delhi: Centre for Air Power Studies, 2018) p.236.
21. A.K. Tiwary, *Indian Air Force in Wars* (New Delhi: Lancer, 2012) pp.295—301.
22. Chordia, *Operation Cactus*, p.235.
23. This figure is a composite from available sources, one of the best being C. V. Gole 'The IAF in 2001' in Vayu Aerospace: II/1994, p.42. However, in the early 1990s, two squadrons converted to the MiG-27, including one from MiG-23BNs. MiG-27 squadrons are Nos. 2, 9, 10, 18, 20, 22, 29 and 222. MiG-23BN squadrons are Nos. 31, 220 and 221. Jaguars operate with Nos. 5, 6, 14, 16 and 27 squadrons.
24. D. Donald & J. Lake, *Encyclopaedia of World Military Aircraft: Volume Two* (London: Aerospace Publishing Ltd., 1994) p.305.
25. R. Rikhye, P. Singh & P. Steinemann, *Fiza'Ya: Psyche of the Pakistan Air Force* (New Delhi: S. A. S., 1991), pp.163–164.
26. Sanjay Badri-Maharaj, *Kargil 1999 – South Asia's First Post-Nuclear Conflict* (Warwick, Helion: 2020), pp.69–71.
27. M Kaiser Tufail, 'Role of the Pakistan Air Force During the Kargil Conflict', *CLAWS Journal* (Summer 2009), pp.104–106.
28. R.K. Simha 'How the IAF Dominated the Skies during Kargil War', *Russia Beyond the Headlines* (26 July 2016). <https://www.rbth.com/blogs/stranger_than_fiction/2016/07/26/how-the-iaf-dominated-the-skies-during-kargil-war_615175>, accessed 26 July 2016.
29. 'India loses two jets', BBC News, <http://news.bbc.co.uk/2/hi/south_asia/354120.stm>, accessed 27 May 1999.
30. Sanjay Badri-Maharaj, *Kargil 1999 – South Asia's First Post-Nuclear Conflict* (Warwick, Helion: 2020) pp.45–51.
31. Badri-Maharaj, *Kargil 1999 – South Asia's First Post-Nuclear Conflict*, pp.45–51.
32. Bharat-rakshak.com, 'The Mirage 2000 in Kargil', (16 October 2009), <http://www.bharat-rakshak.com/IAF/history/kargil/1056-pcamp.html>, accessed 16 October 2009.
33. Bharat-rakshak.com, 'The Mirage 2000 in Kargil.'
34. Bharat-rakshak.com, 'The Mirage 2000 in Kargil.'
35. Bharat-rakshak.com, 'The Mirage 2000 in Kargil.'
36. Bharat-rakshak.com, 'The Mirage 2000 in Kargil.'
37. Shiv Aroor & Rahul Singh, *India's Most Fearless 2* (Gurgaon: Penguin, 2019), pp.xiii–xxii.
38. Shiv Aroor, '80% bombs hit target: IAF gives satellite images to govt as proof of Balakot airstrike', *India Today* (6 March 2019), <https://www.indiatoday.in/india/story/indian-air-force-balakot-airstrike-dossier-satellite-images-bombs-1471355-2019-03-06>, accessed 6 March 2019.
39. Amrita Nayak Dutta, 'IAF's Balakot strikes killed 130-170 Jaish terrorists, claims Italian journalist', *The Print* (8 May 2019), <https://theprint.in/defence/iafs-balakot-strikes-killed-130-170-jaish-terrorists-claims-italian-journalist/232809/>, accessed 8 May 2019.
40. 'Massive cover-up followed Balakot airstrike: Italian journalist to WION', *WION* (8 May 2019), <https://www.wionews.com/south-asia/130-170-terrorists-died-following-balakot-airstrikes-says-award-winning-italian-journalist-217030>, accessed 8 May 2019.
41. Jamshed Khan, Ankit Kumar, Sushant Pathak, 'Balakot Tapes Expose: 4 Pakistani soldiers died in air strikes, confirm locals', *India Today* (11 March 2019), <https://www.indiatoday.in/india/story/balakot-tapes-expose-pakistan-army-deaths-1475497-2019-03-11>, accessed 11 March 2019.
42. 'China, India commence withdrawal of forces from shared border – Chinese Defense Ministry' *TASS* (10 February 2021) <https://tass.com/world/1254813>, accessed 10 February 2021.
43. Jamshed Khan, Ankit Kumar, Sushant Pathak, 'Balakot Tapes Expose: 4 Pakistani soldiers died in air strikes, confirm locals', *India Today* (11 March 2019) <https://www.indiatoday.in/india/story/balakot-tapes-expose-pakistan-army-deaths-1475497-2019-03-11>, accessed 11 March 2019.
44. A.K. Tiwary, *Indian Air Force in Wars* (New Delhi, Lancer: 2012), pp.314–320.

Chapter 2

1. R.G.C. Thomas, 'The Growth of India's Military Powers' in R. Babbage & S. Gordon, eds., *India's Strategic Future* (London: MacMillan, 1992), p.43.
2. R.G.C. Thomas, 'The Growth of India's Military, pp.43–45.
3. *Joint Doctrine Indian Armed Forces* (Headquarters, Integrated Defence Staff, New Delhi: 2017) pp.3–4.
4. *Joint Doctrine Indian Armed Forces*, pp.6–8.
5. *Joint Doctrine Indian Armed Forces*, pp.13–14.
6. *Basic Doctrine of the Indian Air Force 2012* (Indian Air Force: New Delhi, 2012), p.32.
7. *Basic Doctrine of the Indian Air Force 2012*, pp.25–27.
8. *Basic Doctrine of the Indian Air Force 2012*, pp.36.
9. 'Cabinet Committee On Security Reviews Progress In Operationalizing India's Nuclear Doctrine', (4 January 2003) <http://pib.nic.in/archieve/lreleng/lyr2003/rjan2003/04012003/r040120033.html>
10. Shishir Gupta, 'Maritime, air defence theatre commands to be announced by June 2021', *Hindustan Times* (23 March 2021) <https://www.hindustantimes.com/india-news/maritime-air-defence-theatre-commands-to-be-announced-by-june-2021-101616478448044.html>, accessed 23 March 2021.

Chapter 3

1. Bharat-rakshak.com, 'Commands', (10 January 2010), <http://www.bharat-rakshak.com/IAF/units/commands.html#gsc.tab=0>, accessed 10 January 2010.
2. Bharat-rakshak.com, 'Indian Air Force Wings', (10 January 2010), <http://www.bharat-rakshak.com/IAF/units/bases/282-wings.html#gsc.tab=0>, accessed 10 January 2010.
3. Bharat-rakshak.com, 'Forward Base Support Units' (10 January 2010) <http://www.bharat-rakshak.com/IAF/units/bases/284-fbsus.html#gsc.tab=0>, accessed 10 January 2010.
4. Bharat-rakshak.com, 'Air Force Stations', (10 January 2010) <http://www.bharat-rakshak.com/IAF/units/bases/283-air-force-stations.html#gsc.tab=0>, accessed 10 January 2010.
5. Bharat-rakshak.com, 'Indian Air Force Fleet Strength', (6 January 2020), <http://www.bharat-rakshak.com/IAF/units/others/281-fleet.html>, accessed 6 January 2020.
6. <www.Bharat-Rakshak.com> IAF Aircraft Fleet Strength.
7. Bharat-rakshak.com, 'Indian Air Force Fleet Strength', (6 January 2020), <http://www.bharat-rakshak.com/IAF/units/others/281-fleet.html>, accessed 6 January 2020.
8. S. Badri-Maharaj, 'It's Not Just About Combat Aircraft', *Geopolitics* (October 2019), pp.26–30.

9   Badri-Maharaj, 'It's Not Just About Combat Aircraft', pp.26–30.
10 Badri-Maharaj, 'It's Not Just About Combat Aircraft', pp.26–30.
11 Badri-Maharaj, 'It's Not Just About Combat Aircraft', pp.26–30.
12 Badri-Maharaj, 'It's Not Just About Combat Aircraft', pp.26–30.
13 Badri-Maharaj, 'It's Not Just About Combat Aircraft', pp.26–30.
14 Badri-Maharaj, 'It's Not Just About Combat Aircraft', pp.26–30.
15 S. Badri-Maharaj, 'Indian Air Force: Capabilities, Capability Gaps and Options', *Geopolitics* (February 2021) pp.62–66.
16 Badri-Maharaj, 'Indian Air Force: Capabilities, Capability Gaps and Options', pp.62–66.
17 R. Bikhchandani, 'Heron, Searcher, Sea Guardian, SWITCH — the many UAVs that make up India's drone arsenal', *The Print* (6 August, 2021) <https://theprint.in/defence/heron-searcher-sea-guardian-switch-the-many-uavs-that-make-up-indias-drone-arsenal/709670/>, accessed 6 August 2021.
18 Manu Pubby, 'Government approves $400-million plan to procure armed Heron TP drones from Israel', *Economic Times* (14 July 2018) <https://economictimes.indiatimes.com/news/defence/government-approves-400-million-plan-to-procure-armed-heron-tp-drones-from-Israel/articleshow/48906195.cms?utm_source=contentofinterest&utm_medium=text&utm_campaign=cppst>, accessed 14 July 2018.
19 Bharat-rakshak.com, 'Unmanned Aerial Vehicle Squadrons, (6 January 2020), <http://www.bharat-rakshak.com/IAF/units/others/288-uav-squadrons.html#gsc.tab=0>, accessed 6 January 2020.
20 S. Badri-Maharaj, 'Flight Training in the Indian Air Force', *Geopolitics* (October 2019), pp.70–71.
21 Badri-Maharaj, 'Flight Training in the Indian Air Force', pp.70–71.
22 Badri-Maharaj, 'Flight Training in the Indian Air Force', pp.70–71.
23 Badri-Maharaj, 'Flight Training in the Indian Air Force', pp.70–71.
24 Badri-Maharaj, 'Flight Training in the Indian Air Force', pp.70–71.
25 Badri-Maharaj, 'Flight Training in the Indian Air Force', pp.70–71.
26 Badri-Maharaj, 'Flight Training in the Indian Air Force', pp.70–71.
27 Badri-Maharaj, 'Flight Training in the Indian Air Force', pp.70–71.
28 Badri-Maharaj, 'Flight Training in the Indian Air Force', pp.70–71.
29 S. Badri-Maharaj, 'India's Military Special Forces', *Geopolitics* (November 2019) pp. 18–24.
30 Badri-Maharaj, 'India's Military Special Forces', pp. 18–24.
31 Badri-Maharaj, 'India's Military Special Forces', pp. 18–24.

Chapter 4
1   D. Peri, 'CCS okays 83 LCAs worth around ₹47,000 cr. for IAF', *The Hindu* (13 January 2021) <https://www.thehindu.com/news/national/ccs-approves-83-lca-mk-1a-jets-worth-over-48000-crore/article33568729.ece >, accessed 13 January 2021.
2   Anantha Krishna M., 'India's Medium Weight Fighter set to fly into detail design phase', *Onmanorama* (3 February 2020), <https://www.onmanorama.com/news/india/2020/02/03/india-defence-expo-medium-weight-fighter.html>, accessed 3 February 2020.
3   K. Kaushik, 'Tejas done, focus on three other fighter jets: two for IAF, one Navy', *Indian Express* (5 February 2021) <https://indianexpress.com/article/india/aero-india-2021-tejas-done-focus-on-three-other-fighter-jets-two-for-iaf-one-navy-7175125/>, accessed 5 February 2021.
4   'Russia hands over commercial offer of 21 MiG-29 fighters to India', *ANI* (21 July 2021) <https://www.aninews.in/news/world/asia/russia-hands-over-commercial-offer-of-21-mig-29-fighters-to-india20210721060210/>, accessed 21 July 2021.
5   S. Sharma, 'MiG-29 Upgraded: IAF Shares MiG-29 Upgrade Pictures From Deployment In Ladakh; Watch Them Soar', *Republic World* (8 April 2021) <https://www.republicworld.com/india-news/general-news/iaf-shares-mig-29-upgrade-pictures-from-deployment-in-ladakh-watch-them-soar.html>, accessed 8 April 2021.
6   S.A. Philip, 'Why India is set to miss 2021 deadline to upgrade Mirage 2000 fighters', *The Print* (7 October 2021) <https://theprint.in/defence/why-india-is-set-to-miss-2021-deadline-to-upgrade-mirage-2000-fighters/746444/>, accessed 7 October 2021.
7   H. Siddiqui, 'With upgraded avionics, Jaguar still lacks the thrust', *Financial Express* (29 September 2019) <https://www.financialexpress.com/defence/with-upgraded-avionics-jaguar-still-lacks-the-thrust/1721036/>, accessed 29 September 2019.
8   'IAF successfully test fires air-to-air 'Astra' missile from Sukhoi-30 MKI', *Hindustan Times* (19 September 2019) <https://www.hindustantimes.com/india-news/iaf-successfully-test-fires-air-to-air-astra-missile-from-sukhoi-30-mki/story-JZ8ijuSchUOfACTCOpbXSM.html>, accessed 19 September 2019.
9   'India successfully test-fires first indigenous 'Rudram' Anti-Radiation Missile from Sukhoi-30', *Financial Express* (9 October 2020) <https://www.financialexpress.com/defence/india-successfully-test-fires-first-indigenous-rudram-anti-radiation-missile-from-sukhoi-30/2101752/>, accessed 9 October 2020.
10 'India Begins User Trials of Russian Air-to-Air Missiles, Destroys UK-Made Drone in Drill', *Sputnik* (20 June 2019) <https://sputniknews.com/20190620/india-begins-user-trials-of-russian-air-to-air-missiles-destroys-uk-made-drone-in-drill-1075979034.html>, accessed 20 June 2019.
11 H.K. Rout, 'IAF test fires air-to-air missiles ahead of Rafale integration', *New Indian Express* (19 August 2020) <https://www.newindianexpress.com/states/odisha/2020/aug/19/iaf-test-fires-air-to-air-missiles-ahead-of-rafale-integration-2185422.html>, accessed 19 August 2020.
12 S. Badri-Maharaj, 'New Helicopters – Indigenous Progress', *Geopolitics* (April 2020) pp.18-23.
13 Badri-Maharaj, 'New Helicopters – Indigenous Progress', pp.18-23.
14 Badri-Maharaj, 'New Helicopters – Indigenous Progress', pp.18-23.
15 Badri-Maharaj, 'New Helicopters – Indigenous Progress', pp.18-23.
16 Badri-Maharaj, 'New Helicopters – Indigenous Progress', pp.18-23.

Chapter 5
1   J. Baranwal, *SP's Military Yearbook 1992–93* (Guide Publications: New Delhi, 1993), p.SS13.
2   P. Sengupta, 'Arudhra MPR Is EL/M-2084 MMR: Seeing Is Believing', *Trishul-Trident* (4 June 2011) <http://trishul-trident.blogspot.com/2011/06/arudhra-mpr-is-elm-2084-mmr-seeing-is.html>, accessed 4 June 2011.
3   V. Bhatia, 'IAF to deploy new radar near border areas', *Times of India* (2 October 2011) <https://timesofindia.indiatimes.com/city/jaipur/IAF-to-deploy-new-radar-near-border-areas/articleshow/10204373.cms>, accessed 2 October 2011.
4   P. Sengupta, 'IAF's Multi-Phase IACCCS Being Enhanced', *Trishul-Trident* (22 January 2012), <http://trishul-trident.blogspot.com/2012/01/iafs-multi-phase-iacccs-being-enhanced.html>, accessed 22 January 2012.
5   P. Sengupta, 'IAF's Multi-Phase IACCCS Being Enhanced'.
6   P. Sengupta, 'IAF's Multi-Phase IACCCS Being Enhanced'.
7   'Electronic Weapons: Instant Radar', *Strategy Page* (21 February 2009) <https://www.strategypage.com/htmw/htecm/articles/20090221.aspx>, accessed 21 February 2009.
8   P. Sengupta, 'IAF's Multi-Phase IACCCS Being Enhanced', *Trishul-Trident* (22 January 2012) <http://trishul-trident.blogspot.com/2012/01/iafs-multi-phase-iacccs-being-enhanced.html>, accessed 22 January 2012.
9   P. Sengupta, 'IAF's Multi-Phase IACCCS Being Enhanced'.
10 P. Sengupta, 'IAF's Multi-Phase IACCCS Being Enhanced'.
11 P. Sengupta, 'IAF's Multi-Phase IACCCS Being Enhanced'.
12 V. Raghuvanshi, 'Indian Excludes Foreign Vendors From Its Air-Defense Upgrade', *Defense News* (13 June 2016) <https://www.defensenews.com/global/asia-pacific/2016/06/13/indian-excludes-foreign-vendors-from-its-air-defense-upgrade/>, accessed 13 June 2016. See also C. Kumar, 'Six new Akash squadrons to give IAF missile' muscle"
Times of India (17 February 2015), <https://timesofindia.indiatimes.com/india/Six-new-Akash-squadrons-to-give-IAF-missile-muscle/articleshow/46269673.cms>, accessed 17 February 2015.
13 A. Shukla, 'Army orders surface to air missile, making it the first tri-service weapon', *Business Standard* (26 September 2017) <http://www.business-standard.com/article/economy-policy/army-orders-surface-to-air-missile-making-it-the-first-tri-service-weapon-117092500988_1.html>, accessed 26 September 2017.
14 'India acquires Green Pine radars from Israel', *Press Trust of India* (28 June 2002), <https://timesofindia.indiatimes.com/india/India-acquires-Green-Pine-radars-from-Israel/articleshow/14351441.cms>, accessed 28 June 2002.
15 P. Sawhney, 'Decks Cleared For the Contract Signing of S-400 ADMS in December', *Force India* (17 October 2017) <http://forceindia.net/decks-cleared-contract-signing-s-400-adms-december/>, accessed 17 October 2017.
16 'Missile defence shield to be ready in three years: India', *Dawn* (13 December 2007) <https://www.dawn.com/news/280120>, accessed 13 December 2007.
17 S. Jha, 'Hit-to-Kill Successfully Demonstrated By DRDO's PDV Interceptor', *Delhi Defence Review* (25 February 2017) <http://www.delhidefencereview.com/2017/02/25/hit-kill-successfully-demonstrated-drdos-pdv-interceptor/>, accessed 25 February 2017.
18 T.S. Subramanian, 'Smashing hit', *Frontline* (22 December 2007 – 4 Jan 2008) <http://www.frontline.in/static/html/fl2425/stories/20080104242512300.htm>, accessed 4 January 2008.

19 'Interceptor missile scores "direct hit"', *The Hindu* (7 December 2007), <http://www.hindu.com/2007/12/07/stories/2007120761241800.htm>, accessed 7 December 2007.
20 'Interceptor missile test fired successfully', NDTV.com, (6 March 2011), <https://www.ndtv.com/india-news/interceptor-missile-test-fired-successfully-449178>, accessed 6 March 2011.
21 'India proves capability of missile defence system, *The Hindu* (23 November 2012), <https://www.thehindu.com/news/national/india-proves-capability-of-missile-defence-system/article4126430.ece>, accessed 23 November 2012.
22 Y. Mallikarjun, 'Interceptor missile test off Odisha coast fails', *The Hindu* (6 April 2015) <http://www.thehindu.com/news/national/intercepter-missile-test-odisha-coast-wheeler-island-drdo/article7073662.ece?ref=sliderNews>, accessed 6 April 2015.
23 Y. Mallikarjun, 'Upgraded interceptor missile successfully hits virtual target', *The Hindu* (22 November 2015) <http://www.thehindu.com/news/cities/Hyderabad/upgraded-interceptor-missile-successfully-engages-electronically-simulated-target-missile/article7905298.ece>, accessed 22 November 2015.
24 'India successfully test-fires supersonic interceptor missile', *Tribune India* (15 May 2016) <http://www.tribuneindia.com/news/nation/india-successfully-test-fires-supersonic-interceptor-missile/237107.html>, accessed 15 May 2016.
25 T.S. Subramanian, 'Interceptor missile mission a "failure"', *The Hindu* (23 May 2016), <http://www.thehindu.com/news/national/Interceptor-missile-mission-a-%E2%80%98failure%E2%80%99/article14333898.ece>, accessed 8 November 2017.
26 DDR Staff, 'DRDO's AAD Ballistic Missile Defence Interceptor Heads Toward Induction With Latest Test', *Delhi Defence Review* (29 December 2017) <http://www.delhidefencereview.com/2017/12/29/drdos-aad-ballistic-missile-defence-interceptor-heads-toward-induction-with-latest-test/>, accessed 29 December 2017.
27 S. Jha, 'DRDO's AAD Endo-atmospheric Ballistic Missile Interceptor Hits Bullseye', *Delhi Defence Review* (1 March 2017) <http://www.delhidefencereview.com/2017/03/01/drdos-aad-endo-atmospheric-ballistic-missile-interceptor-hits-bullseye/>, accessed 1 March 2017.
28 Jha, 'DRDO's AAD Endo-atmospheric Ballistic Missile Interceptor Hits Bullseye'.
29 Air Marshal, N. Menon, 'Ballistic Missile Defence System for India', *Indian Defence Review Vol 27.3* (July–September 2012) <http://www.indiandefencereview.com/spotlights/ballistic-missile-defence-system-for-india/>, accessed September 2012.
30 'India Successfully Test-Fires New Interceptor Missile', *Outlook India,* (27 April 2014) <https://web.archive.org/web/20140428163936/http://news.outlookindia.com/items.aspx?artid=838755>, accessed 27 April 2014.
31 M. Somasekhar, 'India's "interceptor" may make ballistic missile shield real', *The Hindu Business Line* (12 February 2017) <http://www.thehindubusinessline.com/news/indias-interceptor-may-make-ballistic-missile-shield-real/article9537665.ece>, accessed 12 February 2017.
32 DDR Staff, 'DRDO's AAD Ballistic Missile Defence Interceptor Heads Toward Induction With Latest Test', *Delhi Defence Review* (29 December 2017) <http://www.delhidefencereview.com/2017/12/29/drdos-aad-ballistic-missile-defence-interceptor-heads-toward-induction-with-latest-test/>, accessed 29 December 2017.
33 DDR Staff, 'DRDO's AAD Ballistic Missile Defence Interceptor Heads Toward Induction With Latest Test'.
34 DDR Staff, 'DRDO's AAD Ballistic Missile Defence Interceptor Heads Toward Induction With Latest Test'.
35 DDR Staff, 'DRDO's AAD Ballistic Missile Defence Interceptor Heads Toward Induction With Latest Test'.
36 P. Sengupta, 'AAD Endo-Atmospheric Interceptor Headed For Systems Maturity', *Trishul-Trident Blogspot* (1 January 2018) <http://trishul-trident.blogspot.in/2018/01/aad-endo-atmospheric-interceptor-headed.html>, accessed 1 January 2018.
37 Sengupta, 'AAD Endo-Atmospheric Interceptor Headed For Systems Maturity'.
38 Sengupta, 'AAD Endo-Atmospheric Interceptor Headed For Systems Maturity'.
39 Sengupta, 'AAD Endo-Atmospheric Interceptor Headed For Systems Maturity'.
40 S. Jha, 'Hit-to-Kill Successfully Demonstrated By DRDO's PDV Interceptor', *Delhi Defence Review* (25 February 2017) <http://www.delhidefencereview.com/2017/02/25/hit-kill-successfully-demonstrated-drdos-pdv-interceptor/>, accessed 25 February 2017.
41 Snehesh Alex Philip, 'India's ballistic missile shield ready, IAF & DRDO to seek govt nod to protect Delhi', *The Print* (8 January 2020) <https://theprint.in/defence/indias-ballistic-missile-shield-ready-iaf-drdo-to-seek-govt-nod-to-protect-delhi/345853/>, accessed 8 January 2020.

Chapter 6

1 D. Albright, 'The Shots heard "round the world"', *The Bulletin of the Atomic Scientists* (July/August 1998), pp.20–25.
2 R. Chidambaran, 'Nuclear Dilemma', *India Today*, (30 April 1994), p.50.
3 'The Long Pause 1974–1989', *Nuclear Weapon Archive,* (30 March 2001) <http://nuclearweaponarchive.org/India/IndiaPause.html>, accessed 30 March 2001.
4 R. Chengappa, *Weapons of Peace* (New Delhi: Harper Collins, 2000), p.305.
5 Chengappa, *Weapons of Peace*, p. 304.
6 Chengappa, *Weapons of Peace*, p. 304.
7 G. Kampani, 'New Delhi's Long Nuclear Journey', *International Security*, Vol. 38, No. 4 (Spring 2014), pp.89–92.
8 G. Perkovich, *India's Nuclear Bomb* (London: University of California Press, 1999), p.313.
9 Chengappa, *Weapons of Peace*, pp. 382–385.
10 G. Kampani, 'New Delhi's Long Nuclear Journey', *International Security*, Vol. 38, No. 4 (Spring 2014), p.91.
11 G. Perkovich, *India's Nuclear Bomb: The Impact on Global Proliferation* (London: University of California Press, 1999), p.242.
12 B. Karnad, *Nuclear Weapons and Indian Security* (New Delhi: Macmillan India Ltd, 2002), p.320.
13 Karnad, *Nuclear Weapons and Indian Security*, p.303.
14 G. Kampani, 'New Delhi's Long Nuclear Journey', *International Security*, Vol. 38, No. 4 (Spring 2014), p.91.
15 Kargil Review Committee, *From Surprise to Reckoning*, p. 190. The committee reported that estimates of the 'number of cores/devices/weapons in Pakistan's possession' varied in different intelligence reports and assessments prepared for the government.
16 G. Kampani, 'New Delhi's Long Nuclear Journey', *International Security*, Vol. 38, No. 4 (Spring 2014), pp.91–92.
17 Kampani, 'New Delhi's Long Nuclear Journey', p.327.
18 Kampani, 'New Delhi's Long Nuclear Journey', pp. 382–384.
19 Kampani, 'New Delhi's Long Nuclear Journey', p.102.
20 Kampani, 'New Delhi's Long Nuclear Journey', p.103.
21 'South Asian Nuclear-Weapon-Free Zone' in *The Arms Control Reporter*, 1993, p. 454. B. 176.
22 Sanjay Badri-Maharaj, *Nuclear India – From Reluctance to Triad* (Warwick: Helion, 2021) pp.9–12.
23 Badri-Maharaj, *Nuclear India – From Reluctance to Triad*, pp.9–12.
24 Badri-Maharaj, *Nuclear India – From Reluctance to Triad*, pp.9–12.
25 R. Chengappa, *Weapons of Peace* (New Delhi: Harper Collins, 2000) p. 437.
26 Chengappa, *Weapons of Peace*, pp.103–104.
27 Chengappa, *Weapons of Peace*, pp.103–104.
28 R. Chengappa, *Weapons of Peace* (New Delhi: Harper Collins, 2000) pp.437–438.
29 'Pakistan was to deploy nukes against India during Kargil war', (3 December 2015), <https://m.economictimes.com/news/defence/pakistan-was-to-deploy-nukes-against-india-during-kargil-war/articleshow/50019153.cms>, accessed 3 December 2015.
30 S. Aroor, '10 reasons why the Indian Rafale is evolution itself', *Daily O (4 July 2017)* <http://www.dailyo.in/variety/rafale-aircraft-brahmos-nuclear-defence/story/1/18157.html>, accessed 4 September 2017.
31 M. Tiwari, 'Unprecedented security at Trishul airbase' *Times of India* (3 January 2016) <http://timesofindia.indiatimes.com/city/bareilly/Unprecedented-security-at-Trishul-airbase/articleshow/50430273.cms>, accessed 4 September 2017.
32 'India Plans Next Generation Bombproof Shelter for 108 Fighter Jets', *Sputnik News* (3 July 2017) <https://sputniknews.com/military/201707031055187413-india-shelter-jets/>, accessed 4 September 2017.
33 D. Albright, 'The Shots Heard "Round the World"', *The Bulletin of the Atomic Scientists 4* (July/August 1998), pp.20–25.
34 S. Aroor, '"True" BrahMos Unleashed Today, Next 900-km Weapon', *Livefist Defence* , (11 March 2017), <https://www.livefistdefence.com/2017/03/true-brahmos-unleashed-today-next-1000-km-weapon.html>
35 'India to Join ICBM club soon – Interview with Dr. S. Christopher' *NDTV (*11 July 2015), <http://www.ndtv.com/video/news/news/india-to-join-icbm-club-soon-374683>, accessed 11 July 2015.

36 The only other indication of production levels came in an interview Dr. Saraswat gave in 2012 to *India Today* where he said, 'All I can tell you is that we will produce more than just 1 or 2 missiles a year.' S. Unnithan, 'India has all the building blocks for an anti-satellite capability', *India Today* (27 April 2012) <http://indiatoday.intoday.in/story/agni-v-drdo-chief-dr-vijay-kumar-saraswat-interview/1/186248.html.>, accessed 27 April 2012.
37 'Ballistic and Cruise Missile Threat 2017', *Defense Intelligence Ballistic Missile Analysis Committee* <http://www.nasic.af.mil/Portals/19/images/Fact%20Sheet%20Images/2017%20Ballistic%20and%20Cruise%20Missile%20Threat_Final_small.pdf>, p.25.
38 B. Karnad, *India's Nuclear Policy* (Westport, CT: Praeger Security International, 2008), p.101.
39 Karnad, *India's Nuclear Policy*, pp.101–102.
40 B. Karnad, *Why India is not a Great Power (Yet)* (New Delhi: Oxford University Press, 2015), p.375.
41 'Now, India has a nuclear triad', *The Hindu:* (18 October 2016) <http://www.thehindu.com/news/national/Now-India-has-a-nuclear-triad/article16074127.ece>, accessed 18 October 2016.
42 Ananatha Krishna M.,'K-15 SLBM is a beast with gen-next tech', *New Indian Express* (30 January 2013) <http://www.newindianexpress.com/nation/2013/jan/30/k-15-slbm-is-a-beast-with-gen-next-tech-445756.html>, accessed 30 January 2013.
43 V. Pereira, 'N-capable Arihant submarine successfully test-fires unarmed missile', *Magalorean.com* (26 November 2015) <http://www.mangalorean.com/n-capable-arihant-submarine-successfully-test-fires-unarmed-missile/>, accessed 26 November 2015.
44 A. Panda, 'India Inches Closer to Credible Nuclear Triad With K-4 SLBM Test' *The Diplomat,* (13 May 2014), <http://thediplomat.com/2014/05/india-inches-closer-to-credible-nuclear-triad-with-k-4-slbm-test/>, accessed 13 Mary 2014.
45 Franz-Stefan Gady, 'India Successfully Tests New Ballistic Missile' *The Diplomat, (*22 March 2016) <http://thediplomat.com/2016/03/india-successfully-tests-new-ballistic-missile/>, accessed 22 March 2016.
46 H.K. Rout, 'EXPRESS EXCLUSIVE: Maiden Test of Undersea K-4 Missile From Arihant Submarine', *New Indian Express* (9 April 2016) <http://www.newindianexpress.com/nation/2016/apr/09/EXPRESS-EXCLUSIVE-Maiden-Test-of-Undersea-K-4-Missile-From-Arihant-Submarine-921990.html>, accessed 9 April 2016.
47 'India successfully test-fires 3,500 km nuclear-capable missile K-4', *New Indian Express* (20 January 2020) <https://www.newindianexpress.com/states/odisha/2020/jan/20/india-successfully-test-fires-3500-km-nuclear-capable-missile-k-4-2091838.html>, accessed 20 January 2020.
48 Rajat Pandit, 'Arihant's N-capable missile "ready to roll"', *Times of India (*25 January 2020) <https://timesofindia.indiatimes.com/india/india-successfully-test-fires-k-4-submarine-launched-missile/articleshow/73589861.cms>, accessed 25 January 2020.
49 'India's nuclear triad is complete with INS Arihant ending its first deterrence patrol', *The Hindu* (5 November 2018) <https://www.thehindu.com/news/national/ins-arihant-completes-deterrence-patrol-india-declares-nuclear-traid-operational/article25425436.ece>, accessed 5 November 2018.
50 This is extrapolated on an estimated production of 1-2 missiles per year for the Agni-I and Agni-II post developmental trials and a somewhat higher rate for the Agni-III and IV with Limited Series Production for the Agni-V assumed to have been started. It is probable that each group has 16 launchers and 48 missiles each. This has been gleaned from interviews and from a documentary on Transport Solutions India Limited which noted that a contract for forty-eight trailers—one for each Agni-V missile intended to be produced—was issued in 2015. See 'Make In India— New Deal For Defence—Transport Solutions India, Episode 9, Segment 1,' YouTube video, from 7:20 minutes onwards. <https://www.youtube.com/watch?v=LIaQ3nOGmEI&t=640s> (Accessible only to Indian users or through a VPN). It is also known that at least 16 Agni-V launchers are on order – see *DRDO Annual Report 2016,* p.39.
51 This is based on two reports – the first in DRDO's annual report of 2016 which indicated that 16 Agni-V launchers had been ordered and the second, being a news documentary which noted that 48 trailers (one for each of the Agni-V missiles on order) had been ordered. If one extrapolates that these figures represent a likely average for each Agni missile group, it is therefore probable that India's intended arsenal, in the short term, is likely to be approximately five groups with 48 missiles and 16 launchers each.

Chapter 7

1 'India sets up Integrated Space Cell', *NDTV* (10 June 2008) <https://web.archive.org/web/20080614000739/http://www.ndtv.com/convergence/ndtv/story.aspx?id=NEWEN20080052615>, accessed 10 June 2008.
2 Air Marshal Anil Chopra, 'India's Military Space Program*'*, *South Asia Defence & Strategic Review* (23 March 2017), <http://www.defstrat.com/india%E2%80%99s-military-space-program>, accessed 23 March 2017.
3 Chopra, 'India's Military Space Program*'*.
4 'CARTOSAT-2F', *Spaceflight 101* (13 January 2018), <http://spaceflight101.com/pslv-c40/cartosat-2f/>, accessed 13 January 2018.
5 'CARTOSAT-2F', *Spaceflight 101*.
6 Surendra Singh, 'Military using 13 satellites to keep eye on foes', *Times of India (*26 June 2017) <https://timesofindia.indiatimes.com/india/military-using-13-satellites-to-keep-eye-on-foes/articleshow/59314610.cms>, accessed 26 June 2017.
7 Indranil Roy, 'All You Need To Know About The PDV MK-II: India's Satellite Killer', *Delhi Defence Review* (3 April 2019) <https://delhidefencereview.com/2019/04/03/all-you-need-to-know-about-the-pdv-mk-ii-indias-satellite-killer/>, accessed 3 April 2019.
8 Roy, 'All You Need To Know About The PDV MK-II: India's Satellite Killer'.
9 Roy, 'All You Need To Know About The PDV MK-II: India's Satellite Killer'.
10 Roy, 'All You Need To Know About The PDV MK-II: India's Satellite Killer'.

ABOUT THE AUTHOR

Sanjay Badri-Maharaj, from Trinidad, received his MA and PhD from the Department of War Studies, Kings College London. His thesis was on India's Nuclear Weapons Program. He has written and published extensively, including four books – *The Armageddon Factor: Nuclear Weapons in the India-Pakistan Context* (2000) and *Indian Nuclear Strategy: Confronting the Potential Nuclear Threat from both Pakistan and China* (2018) and a number of titles for Helion's various @War series. He is an attorney-at-law and has served as a consultant to the Ministry of National Security in Trinidad and was a Visiting International Fellow at the Institute for Defence Studies and Analyses in New Delhi.